The Euahlayi Tribe

A Study of Aboriginal Life in Australia

By
K. Langloh Parker

First published in 1905

Published by Left of Brain Books

Copyright © 2023 Left of Brain Books

ISBN 978-1-396-32616-5

First Edition

PUBLISHER'S PREFACE

About the Book

"This is an effort by K. Langloh Parker to describe in formal terms her understanding of Euahlayi society. The resulting ethnography is factual and well written. Parker was obviously also familiar with the anthropological literature. She was hardly the detached observer that modern ethnography demands, however, at that time this methodology had not been invented yet. This is not necessarily a bad thing. As Andrew Lang points out in the introduction, she lived in close contact with aborigines for many years, and as a female she had access to the women of the tribe, a viewpoint for which we have no other source from that time period. "

(Quote from sacred-texts.com)

About the Author

K. Langloh Parker (1856 - 1940)

"K. Langloh Parker (the K. stands for 'Katie') [1856-1940] lived in the Australian outback most of her life, close to the Eulayhi people. The texts, with their sentient animals and mythic transformations, have a sonambulistic and chaotic narrative that mark them as authentic dreamtime lore. The mere fact that she cared to write down these stories places her far ahead of her contemporaries, who barely regarded native Australians as human."

(Quote from sacred-texts.com)

CONTENTS

INTRODUCTION

NO introduction to Mrs. Langloh Parker's book can be more than that superfluous 'bush' which, according to the proverb, good wine does not need. Our knowledge of the life, manners, and customary laws of many Australian tribes has, in recent years, been vastly increased by the admirable works of Mr. Howitt, and of Messrs. Spencer and Gillen. But Mrs. Parker treats of a tribe which, hitherto, has hardly been mentioned by anthropologists, and she has had unexampled opportunities of study. It is hardly possible for a scientific male observer to be intimately familiar with the women and children of a savage tribe. Mrs. Parker, on the other hand, has had, as regards the women and children of the Euahlayi, all the advantages of the squire's wife in a rural neighbourhood, supposing the squire's wife to be an intelligent and sympathetic lady, with a strong taste for the study of folklore and rustic custom. Among the Zulus, we know, it is the elder women who tell the popular tales, so carefully translated and edited by Bishop Colenso. Mrs. Parker has already published two volumes of Euahlayi tales, though I do not know that I have ever seen them cited, except by myself, in anthropo-logical discussion. As they contain many beautiful and romantic touches, and references to the Euahlayi 'All Father,' or paternal 'super man,' Byamee, they may possibly have been regarded as dubious materials, dressed up for the European market. Mrs. Parker's new volume, I hope, will prove that she is a close scientific observer, who must be reckoned with by students. She has not scurried through the region occupied by her tribe, but has had them constantly under her eyes for a number of years.

My own slight share in the book as it stands ought to be mentioned. After reading the original MS., I catechised Mrs. Parker as to her amount of knowledge of the native language; her methods of obtaining information; and the chances that missionary influence had affected the Euahlayi legends and beliefs. I wrote out her answers, and she read and revised what I had written. I also collected many scattered notices of Byamee into the chapter on that being, which Mrs. Parker has read and approved. I introduced a reference to Mr. Howitt's theory of the 'All Father,' and I added some references to other authorities on the Australian tribes. Except for this, and for a very few purely verbal changes in matter of style, Mrs. Parker's original manuscript is untouched by me. It seems necessary to mention these details, as I have, in other works, expressed my own opinions on Australian religion and customary law.[1] These opinions I have not, so to speak, edited into the work of Mrs. Parker. The author herself has remarked that, beginning as a disciple of Mr. Herbert Spencer in regard to the religious ideas of the Australians--according to that writer, mere dread of casual 'spirits'--she was obliged to alter her attitude, in consequence of all that she learned at first hand. She also explains that her tribe are not 'wild blacks,' though, in the absence of missionary influences, they retain their ancient beliefs, at least the old people do; and, in a decadent form, preserve their tribal initiations, or Boorah. How she tested and controlled the evidence of her informants she has herself stated, and I venture to think that she could hardly have made a better use of her opportunities.

In one point there is perhaps, almost unavoidably, a lacuna or gap in her information. The Euahlayi, she says, certainly do not possess the Dieri and Urabunna custom of Pirrauru or Piraunga-ru, by which married , and unmarried men, of the classes men and women which may intermarry, are solemnly allotted to each other as more or less permanent paramours.[1] That custom, for some unknown reason, is confined to certain tribes possessing the two social divisions with the untranslated names Matteri and Kiraru. These tribes range from Lake Eyre southward, perhaps, as far as the sea. Their peculiar custom is unknown to the Euahlayi, but Mrs. Parker does not inform us concerning any recognised licence which may, as is usual, accompany their Boorah assemblies, or their 'harvest home' of gathered grass seed, which she describes.

Any reader of Mrs. Parker's book who has not followed recent anthropological discussions, may need to be apprised of the nature of these controversies, and of the probable light thrown on them by the full description of the Euahlayi tribe. The two chief points in dispute are (1) the nature and origin of the marriage laws of the Australians; and (2) the nature and origin of such among their ideas and practices as may be styled 'religious.' As far as what we commonly call material civilisation is concerned, the natives of the Australian continent are probably the most backward of mankind, having no agriculture, no domestic animals, and no knowledge of metal-working. Their weapons and implements are of wood, stone, and bone, and they have not even the rudest kind of pottery. But though the natives are all, in their natural state, on or about this common low level, their customary laws, ceremonials, and beliefs are rich in variety.

As regards marriage rules they are in several apparently ascending grades of progress. First we have tribes in which each person is born into one or other of two social divisions usually called 'phratries.' Say that the names of the phratries mean Eagle Hawk and Crow. Each born Crow must marry an Eagle Hawk; each born Eagle Hawk must marry a Crow. The names are derived through the mothers. One obvious result is that no two persons, brother and sister maternal, can intermarry; but the rule also excludes from intermarriage great numbers of persons in no way akin to each other by blood, who merely share the common phratry name, Crow or Eagle Hawk.

In each phratry are smaller sets of persons, each set distinguished by the name of some animal or other natural object, their 'totem.' The same totem is never found in both phratries. Thus a person marrying out of his or her phratry, as all must do, necessarily marries out of his or her totem.

The same arrangements exist among tribes which derive phratry and totem names through the father.

This derivation of names and descent through the father is regarded by almost all students, and by Mr. J. G. Frazer, in one passage of his latest study of the subject, as a great step in progress.[2] The obvious result of paternal descent is to make totem communities or kins local. In any district most of the people will be of the same paternal totem name-say, Grub, Iguana, Emu, or what not. just so, in Glencoe of old, most of the people were Maclans; in Appin most were Stewarts; in South Argyll Campbells, and so on.

The totem kins are thus, with paternal descent, united both by supposed blood ties in the totem kin, and by associations of locality. This is certainly a step in social progress.

But while Mr. Frazer, with almost all inquirers, acknowledges this, ten pages later in his essay he no longer considers the descent of the totem in the paternal line as necessarily 'a step in progress' from descent in the maternal line. 'The common assumption that inheritance of the totem through the mother always preceded inheritance of it through the father need not hold good,'[3] he remarks.

Thus it appears that a tribe has not necessarily made 'a great step in progress,' because it reckons descent of the totem on the male side. If this be so, we cannot so easily decide as to which tribe is socially advanced and which is not.

In any case, however, there is a test of social advance. There is an acknowledged advance when a tribe is divided into, not two, but four or eight divisions, which may not intermarry.[4] The Euahlayi have four such divisions. In each of their intermarrying phratries are two 'Matrimonial Classes,' each with its name, and these are so constituted that a member of the elder generation can never marry a member of the succeeding generation. This rule prevents, of course, marriage between parent and child, but such marriages never do occur in the pristine tribes of the Darling river which have no such classes. The four-class arrangement excludes from intermarriage all persons, whether parents and children or not, who bear the same class name, say Hippai.

Among the central and northern tribes, from the Arunta of the Macdonnell hills to the Gulf of Carpentaria, the eight-class rule exists, and it is, confessedly, the most advanced of all.

In this respect, then, the Arunta of the centre of Australia are certainly more advanced than the Euahlayi. The Arunta have eight, not four, intermarrying classes. In the matter of rites and ceremonies, too, they are, in the opinion of Messrs. Spencer and Gillen, more advanced than, say, the Euahlayi. They practise universal 'subincision' of the males, and circumcision, in place of the more primitive knocking out of the front teeth. Their ceremonies are very prolonged: in Messrs. Spencer and Gillen's experience, rites lasted for four months during a great tribal gathering. That the Arunta could provide supplies for so prolonged and large an assembly, argues high organisation, or a region well found in natural edible objects. Yet the region is arid and barren, so the organisation is very high. For all these reasons, even if we do not regard paternal descent of the totem as a step in progress from maternal descent, the Arunta seem greatly advanced in social conditions.

Yet they are said to lack entirely that belief in a moral and kindly 'All Father,' such as Byamee, which Mrs. Parker describes as potent among the less advanced Euahlayi, and which Mr. Howitt has found among non-coastal tribes of the south-east, with female descent of the totem, but without matrimonial classes- that is, among the most primitive tribes of all.

Here occurs a remarkable difficulty. Mr. Howitt asserts, with Mr. Frazer's concurrence, that (in Mr. Frazer's words) 'the same regions in which the germs of religion begin to appear have also made some progress towards a higher form of social and family life.'[5] But the social advance from maternal to paternal descent of the totem, we have seen, is not necessarily an advance at all, in Mr. Frazer's opinion.[6] The Arunta, for example, he thinks, never recognised female descent of the totem. They have never recognised, indeed, he thinks, any hereditary descent of the totem, though in all other respects, as in hereditary magistracies, and inheritance of the right to practise the father's totemic ritual, they do reckon in the male line. By such advantage, however it was acquired, they are more progressive than, say, the Euahlayi. But, progressive as they are, they have not, like the more pristine tribes of the south-east, developed 'the germs of religion,' the belief in a benevolent or ruling 'All Father.' Unlike the tribes of the south-east, they have co-operative totemic magic. Each totem community does magic for its totem, as part of the food supply of the united tribe. But the tribe, though so solidaire, and with its eight classes and hereditary magistracies so advanced, has developed no germs of religion at all. Arunta progress has thus been singularly unequal.

The germs of religion are spoken of as the results of social advance, but, while so prominent in social advance, the Arunta have no trace of religion. The tribes northward from them to the sea are also very advanced socially, but (with one known exception not alluded to by Mr. Frazer) have no 'All Father,' no germ of religion.

From this fact, if correctly reported, it is obvious that social progress is not the cause, nor the necessary concomitant, of advance in religious ideas.

Again, the influence of the sea, in causing a 'heavier rainfall, a more abundant vegetation, and a more plentiful supply of food,' with an easier and more reflective life than that of 'the arid wilderness of the interior,' cannot be, as is alleged, the cause of the germs of religion.[7] If this were the case, the coastal tribes of the Gulf of Carpentaria and of the north generally would have developed the All Father belief. Yet, in spite of their coastal environment, and richer existence, and social advance, the northern coastal tribes are not credited with the belief in the All Father. Meanwhile tribes with no matrimonial classes, and with female descent of the totem-tribes dwelling from five to seven hundred miles away from the southern sea-do possess the All Father belief as far north as Central Queensland, no less than did the almost or quite extinct tribes of the south coast, who had made what is (or is not) 'the great step in progress' of paternal descent of the totem.

Again, arid and barren as is the central region tenanted by the Arunta, it seems to permit or encourage philosophic reflection, for their theory of evolution is remarkably coherent and ingenious. The theory of evolution implies as much reflection as that of creation! Their magic for the behoof of edible objects is attributed to the suddenness of their first rains,[8] and the consequent outburst of life, which the natives attribute to their own magical success. But rainmaking magic, as Mrs. Langloh Parker shows, is practised with sometimes amazing success among the Euahlayi, who work no magic at all for their totems. Their magic, if it brings rain, benefits their totems at large, but for each totem in particular, no Euahlayi totem kin does magic.

Again, agricultural magic has been, and indeed is, practised in Europe, in conditions of climate unlike those of the Arunta; and totemic magic is freely practised in North America, in climatic conditions dissimilar from those of Central Australia.

For all these reasons I must confess that I do not follow the logic of the philosophy which makes social advance the cause of the belief in the All Father, and coastal rains the cause of social advance. The Arunta have the social advance, the eight classes, the relatively high organisation; but they have neither the climatic conditions supposed to produce the advance, nor the religion which the advance is supposed to produce. The northern coastal tribes, again, have the desired climatic conditions, and the social advance, but they have not the germs of religion found in many far inland southern tribes, like the Euahlayi, whose social progress is extremely moderate. We thus find, from the northern coast to the centre, one supposed result of coastal conditions, namely, social progress, but not the other supposed result of coastal conditions, namely, the All Father belief. I do not say that it does not exist, for it is a secret belief, but it is not reported by Messrs. Spencer and Gillen. On the other hand, among tribes of the south-east very far from the coast, we find the lowest grades of social progress, but we also find the All Father belief. I am ready, of course, to believe that good conditions of life beget progress, social and religious, as a general rule. But other causes exist; speculation anywhere may take crudely scientific rather than crudely religious lines.

Especially the belief in ancestral spirits may check or nullify the belief in a remote All Father. We see this among the Zulus, where spirits entirely dominate religion, and the All Father is, at most, the shadow of a name, Unkulunkulu. We may detect the same influence among the northern tribes of Australia, where ancestral spirits dominate thought and society, though they receive no sacrifice or prayer. Meanwhile, if we accept Mrs. Parker's evidence, among the Euahlayi ancestral spirits are of no account in religion) while the All Father is obeyed, and, on some occasions, is addressed in prayer; and may even cause rain, if property approached by a human spirit which has just entered his mansions. Clearly, climatic causes and natural environment are not the only factors in producing and directing the speculative ideas of men in early society.

We must also remember that the neighbours of the Arunta, northwards, who share certain peculiar Arunta ideas, possess, beyond all doubt, either the earliest germs of belief in the All Father, or that belief in a decadent condition of survival. This is quite certain; for, whereas the Arunta laugh at all inquiries as to what went before the 'Alcheringa,' or mythic age of evolution, the Kaitish, according to Messrs. Spencer and Gillen, aver that an anthropomorphic being, who dwells above the sky, and is named Atnatu, first created himself, and then 'made the Alcheringa,'--the mythic age of primal evolution. Of mankind, some, in Kaitish opinion, were evolved; of others Atnatu is the father. He expelled men to earth from his heaven for neglect of his ceremonies, but he provided them with weapons and all that they possess. He is not très ferrè sur la morale: he has made no moral laws, but his ritual laws, as to circumcision and the whirling of the bull-roarer, must be observed as strictly as the ritual laws of Byamee of the Euahlayi. In this sense of obedience due to a heavenly father who begat men, or some of them, punished them, and started them on their terrene career, laying down ceremonial rules, we have certainly 'the germs of religion' in a central tribe cognate to the Arunta.

Mr. Frazer detects only two traces of religion in the centre, omitting the Kaitish Atnatu,[9] but I am unable to see how the religious aspect of Atnatu, non-moral as it is, can be overlooked. He is the father of part of the tribe, and all are bound to, observe his ceremonial rules. He accounts for the beginning of the beginning; he is the cause of the Alcheringa; men owe duties to him. We do not know whether he was once as potent in their hearts, and as moral as Byamee, but has dégringolé under Arunta philosophic influences; or whether Byamee is a more highly evolved form of Atnatu. But it is quite certain that the Kaitish, in a region as far almost from the north sea as that of the Arunta, and further from southern coastal influences than the Arunta, have a modified belief in the All Father. How are we to account for this on the philosophic hypothesis of Oceanus as the father of all the gods; of coastal influences producing a richer life, and causing both social and religious progress?

Another difficulty is that while the Arunta, with no religion, and the Kaitish, with the Atnatu belief, are socially advanced in organisation (whether we reckon male descent of the totem 'a great step in progress,' or an accident), they are yet supposed by Mr. Frazer to be, in one respect, the least advanced, the most primitive, of known human beings. The reason is this: the Arunta do not recognise the processes of sexual union as the cause of the production of children. Sexual acts, they say, merely prepare women for the reception of original ancestral spirits, which enter into them, and are reincarnated and brought to the birth.

If the women cannot accept the spirits without being 'prepared' by sexual union, then sexual union plays a physical part in the generation of a spirit incarnated, a fact which all believers in the human soul are as ready as the Arunta to admit. If the Arunta recognise the prior necessity of ' preparation,' then they are not so ignorant as they are thought to be; and their view is produced, not so much by stark ignorance, as by their philosophy of the eternal reincarnation of primal human spirits. The Arunta philosophers, in fact, seem to concentrate their speculation on a point which puzzled Mr. Shandy. How does the animating principle, or soul, regarded as immaterial, clothe itself in flesh? Material acts cannot effect the incarnation of a spirit. Therefore, the spirit enters women from without, and is not the direct result of human action.

The south-eastern tribes, with female descent of the totem, and with no belief in the universal and constant reincarnation of ancestral spirits, take the Æschylean view, according to Mr. Howitt, that the male is the sole originating cause of children, while the female is only the recipient and 'nurse.' These tribes, socially less advanced than the Arunta, have not the Arunta nescience of the facts of procreation, a nescience which I regard as merely the consequence and corollary of the Arunta philosophy of reincarnation. Each Arunta child, by that philosophy, has been in being since the Alcheringa: his mother of the moment only reproduces him, after 'preparation.' He is not a new thing; he is as old as the development of organic forms. This is the Arunta belief, and I must reckon it as not more primitive than the peculiar philosophy of reincarnation of ancestral spirits. Certainly such an elaborate philosophy manifestly cannot be primitive. It is, however, the philosophy of the tribes from the Urabunna, on Lake Eyre (with female descent of the totem), to the most northerly tribes, with male descent.

But among none of these tribes has the philosophy that extraordinary effect on totemic institutions which, by a peculiar and isolated addition, it possesses among the septs of the Arunta nation, and in a limited way among the Kaitish.

Among all tribes except these the child inherits its totem: from the mother, among the Urabunna; from the father in the northern peoples. But, among the Arunta and Kaitish, the totem is not inherited from either parent. According to the belief of these tribes, in every district there is a place where the first human ancestors--in each case all of one totem, whichsoever that totem, in each case, might happen to be--died, 'went under the earth.' Rocks or trees arose to mark such spots. These places are haunted by the spirits of the dead ancestors; here they are all Grubs, there all Eagle Hawks, or all Iguanas, or all Emus, or all Cats. Or as in these sites the ancestors left each his own sacred stone, churinga nanja, with archaic patterns inscribed on it, patterns now fancifully interpreted as totemic inscriptions. Such stones are especially haunted by the ancestral souls, all desiring reincarnation.

When a woman becomes aware of the life of the child she bears, among the Arunta and Kaitish, she supposes that a local spirit of the local totem has entered her, and her child's totem is therefore the totem of that locality, whatever other totems she and her husbands may own. The stone amulet of the ancestral spirit, who is the child, is sought; if it cannot be found at the spot, a wooden churinga is made to represent it, and it is kept carefully in a sacred storehouse.

Even in the centre and north, where the belief in reincarnation prevails, this odd manner of acquiring totems is only practised by the Arunta tribes and the Kaitish, and only among them are the inscribed stones known to exist as favoured haunts of ancestral spirits desiring incarnation. The other northern tribes believe in reincarnation, but not in the haunted sacred stones, which they do not, north of the Worgaia, possess; nor do they derive totems from locality, but, as usual, by inheritance.

It thus appears that these Arunta sacred stones are an inseparable accident of the Arunta method of acquiring the totem. How they and the faith in them cause that method is not obvious, but the two things-the haunted sacred stone, and the local source of totems--are inseparable--that is, the former never is found apart from the latter. Now such stones, with the sense and usage attached to them, cannot well be primitive. They are the result of the peculiar and strictly isolated Arunta custom and belief, which gives to each man and woman one of these stones, the property of himself or herself, since the mythical age, through all reincarnations.

One cannot see how such an unique custom and belief, associated with objects of art, can be reckoned primitive. Yet, where such stones do not exist, the usage of acquiring totems by locality does not exist; even where the belief in reincarnation and in local centres haunted by totemic spirits is found in North Australia.[10]

On these grounds it appears that the hereditary totem is the earlier, and that the Arunta usage is the result of the special and inseparable superstition about the sacred stones. It may be a relatively recent complication of and addition to the theory of reincarnation. Meanwhile, the belief and usage produce an unique effect. The Arunta and Kaitish, we saw, are so advanced socially that they possess not two, or four, but eight matrimonial classes. The tribe is divided into two sets of four classes each, and no person in A division (nameless) of four classes may marry another person of any one of these four, but must marry a person of a given class among the four in B division (nameless). The succession to the class is hereditary in the mate line. But any person among the Arunta, contrary to universal custom elsewhere, may marry another person of his or her own totem, if that person be in the right class of the opposite division. Nowhere else can a person of division A and totem Grub find a Grub to marry in the opposite division B. But this is possible among the Arunta and Kaitish, because their totems are acquired by pure accident, are not hereditary, and all totems exist, or may exist, in division A and also in division B.

Mr. Frazer argues that the Arunta is the earlier state of affairs. He supposes that men acquired their totems, at first, by local accident, before they had laid any restrictions on marriage. Later, they divided their tribe, first into two, then into four, then into eight classes; and every one had to marry out of his class, or set of classes. All other known tribes introduced these restrictions after totems had been made hereditary. On passing the restrictive marriage law, they merely drafted people of one set of hereditary totems into one division, all the other totem kins into the other division. But the Arunta had not made totems hereditary, but accidental, so all the children of one crowd of mothers were placed in division A, all other children in division B. The mothers in each division would have children of all the totems, and thus the same totems now appeared in both of the exogamous divisions. If a man married into his lawful opposite class, the fact that the woman was of the same totem made no difference.

I have offered quite an opposite explanation. Arunta totems were, originally, hereditary among the Arunta, as everywhere else, and no totem occurred in both exogamous divisions. The same totems, later, got into both divisions as the result of the later and isolated belief in reincarnation plus the sacred haunted stones. That superstition has left the Kaitish practice of marriage still almost untouched. A Kaitish may, like an Arunta, marry a woman of his own totem, but he scarcely ever does so. The old prohibition, extinct in law, persists in custom; unless we say that the Kaitish are now merely imitating the usual practice of the rest of the totemic races of the world.

Moreover, even among the Arunta, certain totems greatly preponderate in each of the two exogamous intermarrying divisions of the tribe. This must be because the present practice has not yet quite upset the ancient usage, by which no totem ever occurred in both divisions. There is even an Arunta myth asserting that this was so, but it is, of course, of no historical value as evidence. Here it is proper to give Mr. Frazer's contrary theory in his own words:--

'This [Arunta] mode of determining the totem has all the appearance of extreme antiquity. For it ignores altogether the intercourse of the sexes as the cause of offspring, and further, it ignores the tie of blood on the maternal as well as the paternal side, substituting for it a purely local bond, since the members of a totem stock are merely those who gave the first sign of life in the womb at one or other of certain definite spots. This form of totemism, which may be called conceptional or local to distinguish it from hereditary totemism, may with great probability be regarded as the most primitive known to exist at the present day, since it seems to date from a time when blood relationship was not yet recognised, and when even the idea of paternity had not yet presented itself to the savage mind. Moreover, it is hardly possible that this peculiar form of local totemism, with its implied ignorance of such a thing as paternity at all, could be derived from hereditary totemism, whereas it is easy to understand how hereditary totemism, either in the paternal or in the maternal line, could be derived from it. Indeed, among the Umbaia and Gnanji tribes we can see at the present day how the change from local to hereditary totemism has been effected. These tribes, like the Arunta and Kaitish, believe that conception is caused by the entrance into a woman of a spirit who has lived in its disembodied state, along with other spirits of the same totem, at any one of a number of totem centres scattered over the country; but, unlike the Arunta and Kaitish, they almost always assign the father's totem to the child, even though the infant may have given the first sign of life at a place haunted by spirits of a different totem. For example, the wife of a snake man may first feel her womb quickened at a tree haunted by spirits of goshawk people; yet the child will not be a goshawk but a snake, like its father. The theory by which the Umbaia and Gnanji reconcile these apparently inconsistent beliefs is that a spirit of the husband's totem follows the wife and enters into her wherever an opportunity offers, whereas spirits of other totems would not think of doing so. In the example supposed, a snake spirit is thought to have followed up the wife of the snake man and entered into her at the tree haunted by goshawk spirits, while the goshawk spirits would refuse to trespass, so to say, on a snake preserve by quartering themselves in the wife of a snake man. This theory clearly marks a transition from local to hereditary totemism in the paternal line. And precisely the same theory could, mutatis mutandis, be employed to effect a change from local to hereditary totemism in the maternal line; it would only be necessary to suppose that a pregnant woman is always followed by a spirit of her own totem, which sooner or later effects a lodgement in her body. For example, a pregnant woman of the bee totem would always be followed by a bee spirit, which would enter into her wherever and whenever she felt her womb quickened, and so the child would be born of her own bee totem. Thus the local form of totemism, which obtains among the Arunta and Kaitish tribes, is older than the hereditary form, which is the ordinary type of totemism in Australia and elsewhere, first, because it rests on far more archaic conceptions of society and of life; and, secondly, because both the hereditary kinds of totemism, the paternal and the maternal, can be derived from it, whereas it can hardly be derived from either of them.'

This argument appears to take for granted that the conception of primal ancestral spirits, perpetually reincarnated, is primitive. But, in fact, we seem to know it, among Australian tribes, only in these which have advanced to the possession of eight classes, and have made 'the great step in progress' (if it is a great step), of descent of the totem in the paternal line. The Urabunna, with female descent of the totem, have, it is true, the belief in reincarnation. But they intermarry with the Arunta, borrow their sacred stones, and practise the same advanced rites and ceremonies. The idea may thus have been borrowed. On the other hand, the more pristine tribes of the south-east, with two or four exogamous divisions, and with female descent of the totem, have no known trace of the doctrine of reincarnation (except as displayed by the Euahlayi), and have no doubt that the father is the cause of procreation, save in the case of the Euahlayi, who believe that the Moon and the Crow 'make' the new children.

It would thus appear that the central and northern belief in perpetual reincarnation of primal spirits is not primitive, yet the Arunta method of acquiring totems does not exist save by grace of this belief, plus the isolated belief in primal sacred stones.

I am obliged to differ from Mr. Frazer when he says that 'it is easy to see how hereditary totemism, either in the paternal or in the maternal line, would be derived from' the Arunta belief and practice, whereas 'it is hardly possible that this peculiar form of local totemism [Arunta], with its implied ignorance of such a thing as paternity at all, could be derived from hereditary totemism.'

I do not know whether the other northern tribes share the Arunta nescience of procreation, or not. Whether they do or do not, it was as easy for them to e plain all difficulties by a reconciling myth-a spirit of the husband's totem follows his wife-as for a white savant to frame an hypothesis. The Urabunna, with female descent of the totem, have quite another myth-to reconcile everything.

Nothing can be more easy. Supposing the Arunta to have begun, as in my theory, with hereditary totemism, the rise of their isolated belief in spirit-haunted sacred stones, encroached on and destroyed the hereditary character of their totemism. The belief in churinga nanja is an isolated freak, but it has done its work, while leaving traces of an earlier state of things, as we have shown, both among the Kaitish and Arunta.

If I am right in differing from such a master of many legions as the learned author of The Golden Bough, the irreligion of the Arunta and northern tribes (if these be really without religion) is the result of their form of speculation, wholly occupied by the idea of reincarnation, while the Arunta form of totemism is the consequence of an isolated fantasy about their peculiar sacred stones. Meanwhile the Euahlayi, as Mrs. Parker proves, entertain, in a limited way, not elsewhere recorded in Australia, the belief in the reincarnation of the souls of uninitiated young people. They also, like the Arunta, recognise haunted trees and rocks, but the haunting spirits do not desire reincarnation, and are not ancestral. Spirits of the dead go to one or other abode of souls, to Baiame, or far from his presence to a place of pain. So limited is human fancy, that here, as in Beckford's picture of hell in Vathek, each spirit eternally presses his hand against his side. Were this a Christian doctrine, the Euahlayi would be said to have borrowed it, but few will accuse them of plagiarising from Beckford. These myths, like all myths, are not consistent. Baiame may change a soul into a bird.

We may ask whether, with their limited belief in reincarnation, and with their haunted Minggah trees and rocks, the Euahlayi have set up a creed which might possibly develop into the northern faith, or whether they once held the northern faith, and have almost emerged from it. Without further information about intermediate tribes and their ideas on these matters, the question cannot be answered. We are also without data as to whether the nearly extinct southern coastal tribes evolved the All Father belief, and transmitted it to the Euahlayi, to some Queensland tribe, with their Mulkari, and even to the Kaitish, or whether the faith has been independently developed among the tribes with no matrimonial classes and the others. Conjecture is at present useless.

In one respect a discovery of Mrs. Parker's is unfavourable to my theories. In The Secret of the Totem have shown that, when the names of the phratry divisions of the tribes can be interpreted, they prove to be names of animals, and I have shown how this may have come to be the case. But among the Euahlayi the phratry names mean 'light blood' and 'dark blood.' This, prima facie, seems to favour the theory of the Rev. Mr. Mathews, in his Eagle Hawk and Crow, that two peoples, lighter and darker, after an age of war, made connubium and marriage treaty, whence came the phratries. The same author might urge, if he pleased, that Eagle Hawk (about the colour of the peregrine) was chosen to represent 'light,' and Crow to represent 'dark'; while the phratry animals, White and Black Cockatoo, were selected, elsewhere, to represent the same contrast. But we need more information as to the meanings of other phratry names which have defied translation.

In many other things, as in the account of the yunbeai of the Euahlayi, their mode of removing the tabu on the totem in food, their magic, their 'multiplex totems,' their methods of hunting, their initiatory ceremonies, their highly moral lullabies, and the whole of their kindly life, Mrs. Parker's book appears to deserve a welcome from the few who care to study the ways of early men, 'the pit whence we were dug.' The Euahlayi are a sympathetic people, and have found a sympathetic chronicler.

A. LANG.

INTRODUCTORY

THE following pages are intended as a contribution to the study of the manners, customs, beliefs, and legends of the Aborigines of Australia. The area of my observation is mainly limited to the region occupied by the Euahlayi tribe of north-western New South Wales, who for twenty years were my neighbours on the Narran River. I have been acquainted since childhood with the natives, first in southern South Australia; next on my father's station on the Darling River, where I was saved by a native girl, when my sisters were drowned while bathing. I was intimate with the dispositions of the blacks, and was on friendly terms with them, before I began a regular attempt to inquire into their folk-lore and customary laws, at my husband's station on the Narran, due north of the Barwon River, the great affluent of the Murray River.

My tribe is a neighbour of that mentioned by Mr. Howitt as the 'Wollaroi,' 'Yualloroi,' or 'Yualaroi.'[11] I spell the tribal name 'Euahlayi'; the accent is on the second syllable--'You-ahl-ayi'; and the name is derived from the tribal word for the negative: Euahl, or Youal, 'No,' as in the case of the Kamilaroi (Kamil, 'No'), and many other tribes.

Mr. Howitt regards these tribes as on the limits of what he calls the 'Four Sub-Class' system. The people, that is to say, have not only the division into two 'phratries,' or 'exogamous moieties,' intermarrying, but also the four 'Matrimonial Classes' further regulating marriage. These classes bear the Kamilaroi names, of unknown meaning, Ipai, Kumbo, Murri, and Kubbi; but the names of the two main divisions, or phratries, are not those of the Kamilaroi--Dilbi and Kupathin.

The Euahlayi language, or dialect, is not identical with that of the great Kamilaroi tribe to their south-east, but is clearly allied with it, many names of animals being the same in both tongues. A few names of animals are shared with the Wirádjuri speech, as Mullian, Eagle Hawk; Pelican, Goolayyahlee (Wirádjuri, Gulaiguli). The term for the being called 'The All Father' by Mr. Howitt is also the term used by the Wirádjuri and Kamilaroi, 'Baiame' or 'Byamee.' The Euahlayi, however, possess myths, beliefs, and usages not recorded as extant among the Kamilaroi, but rather forming a link with the ideas of peoples dwelling much further west, such as the tribes, on Lake Eyre, and the southernmost Arunta of the centre. Thus, there is a limited and modified shape of the central and northern belief in reincarnation, and there is a great development of what are called by Mr. Howitt ' sub-totems,' which have been found most in a region of Northern Victoria, to the south of the Euahlayi. There is a belief in spirit-haunted trees, as among the Arunta, and there is a form of the Arunta myth of the 'Dream Time,' the age of pristine evolution.

The Euahlayi thus present a mixture of ideas and usages which appears to be somewhat peculiar and deserving of closer study than it has received. Mr. Howitt himself refers to the tribe very seldom. It will be asked, 'How far have the Euahlayi been brought under the influence of missionaries, and of European ideas in general?'

The nearest missionary settlement was founded after we settled among the Euahlayi, and was distant about one hundred miles, at Brewarrina. None of my native informants had been at any time, to my knowledge, under the influence of missionaries. They all wore shirts, and almost all of them trousers, on occasion; and all, except the old men, my chief sources, were employed by white settlers. We conversed in a kind of lingua franca. An informant, say Peter, would try to express himself in English, when he thought that I was not successful in following him in his own tongue. With Paddy, who had no English but a curse, I used two native women, one old, one younger, as interpreters, checking each other alternately. The younger natives themselves had lost the sense of some of the native words used by their elders, but the middle-aged interpreters were usually adequate. Occasionally there were disputes on linguistic points, when Paddy, a man already grey in 1845, would march off the scene, and need to be reconciled. They were on very good terms with me. They would exchange gifts with me: I might receive a carved weapon, and one of them some tobacco. The giving was not all on my side, by any means.

My anthropological reading was scanty, but I was well acquainted with and believed in Mr. Herbert Spencer's 'Ghost theory' of the origin of religion in the worship of ancestral spirits. What I learned from the natives surprised me, and shook my faith in Mr. Spencer's theory, with which it seemed incompatible.

In hearing the old blacks tell their legends you notice a great difference between them as raconteurs--some tell the bare plot or feature of the legend, others give descriptive touches all through. If they are strangers to their audience, they get it over as quickly as possible in a half-contemptuous way, as if saying, 'What do you want to know such rubbish for?' But if they know you well, and know you really are interested, then they tell you the stories as they would tell them to one another, giving them a new life and adding considerably to their poetical expression.

THE ALL FATHER, BYAMEE

As throughout the chapters on the customary laws, mysteries, and legends of the Euahlayi, there occur frequent mentions of a superhuman though anthropo-morphic being named Byamee (in Kamilaroi and Wirádjuri 'Baiame'), it is necessary to give a preliminary account of the beliefs entertained concerning him. The name Byamee (usually spelled Baiame) occurs in Euahlayi, Kamilaroi, and Wirádjuri; 'the Wirádjuri language is spoken over a greater extent of territory than any other tongue in New South Wales.'[12] The word occurs in the Rev. Mr. Ridley's Gurre Kamilaroi, an illustrated manual of Biblical instruction for the education of the Kamilaroi: Mr. Ridley translated our 'God' by 'Baiame.' He supposed that native term, which he found and did not introduce, to be a derivative from the verb baia, or biai, 'to make.' Literally, however, at least in Euahlayi, the word byamee means 'great one.' In its sense as the name of the All Father it is not supposed to be used by women or by the uninitiated. If it is necessary to speak to them of Byamee, he is called Boyjerh, which means Father, just as in the Theddora tribe the women speak of Darramulun as Papang, 'Father.'[13] Among the Euahlayi both women and the uninitiated use byamee, the adjective for 'great,' in ordinary talk, though the more usual adjective answering to 'great' is boorool, which occurs in Kamilaroi as well as in Euahlayi. The verb baia or biai, to make or shape, whence Mr. Ridley derived Baiame, is not known to me in Euahlayi. Wirádjuri has bai, a footmark, and Byamee left footmarks on the rocks, but that is probably a chance coincidence.

I was first told of Byamee, in whispers, by a very old native, Yudtha Dulleebah (Bald Head), said to have been already grey haired when Sir Thomas Mitchell discovered the Narran in 1846. My informant said that he was instructed as to Byamee in his first Boorah, or initiation. If he was early grey, say at thirty, in 1846, that takes his initiation back to 1830, when, as a matter of fact, we have contemporary evidence to the belief in Byamee, who is not of missionary importation, though after 1856 Christian ideas may, through Mr. Ridley's book, have been attached to his name by educated Kamilaroi. But he was a worshipful being, revealed in the mysteries, long before missionaries came, as all my informants aver.

There has, indeed, been much dispute as to whether the Aborigines of Australia have any idea, or germ of an idea, of a God; anything more than vague beliefs about unattached spirits, mainly mischievous, who might be propitiated or scared away. Mr. Huxley maintained this view, as did Mr. Herbert Spencer.[14] Both of these authors, who have great influence on popular opinion, omitted to notice the contradictory statement of Waitz, published in 1872. He credited the natives, in some regions, with belief in, and dances performed in honour of, a 'Good Being,' and denied that the belief and rites were the result of European influence.[15] Mr. Tylor, admitting to some extent that the belief now exists, attributed it in part to the influence of missionaries and of white settlers.[16] 'Baiame,' he held, was a word of missionary manufacture, introduced about 1830-1840. This opinion was controverted by Mr. Lang,[17] and by Mr. N. W. Thomas. Mr. Thomas[18] has produced the evidence of Henderson, writing in 1829-1830, for the belief in 'Piame' or Byamee, or Baiame.[19]

In 1904 Mr. Howitt gave a great mass of evidence for the belief in what he calls an 'All Father': in many dialects styled by various names meaning 'Our Father,' dwelling in or above the sky, and often receiving the souls of blacks who have been 'good.' These ideas are not derived, Mr. Howitt holds, from Europeans, or developed out of ancestor-worship, which does not exist in the tribes. The belief is concealed from women, but communicated to lads at their initiation.[20] The belief, in favourable circumstances, might develop, Mr. Howitt thinks, into what he speaks of as a 'religion,' a 'recognised religion.' Without asking how 'a recognised religion' is to be defined, I shall merely tell what I have gathered as to the belief in Byamee among the Euahlayi.

It may seem strange that I should know anything about a belief carefully kept from women, but I have even been privileged to hear 'Byamee's Song,' which only the fully initiated may sing; an old black, as will later appear, did chant this old lay, now no longer understood, to myself and my husband. Moreover, the women of the Euahlayi have some knowledge of, and some means of, mystic access to Byamee, though they call him by another name.

Byamee, in the first place, is to the Euahlayi what the 'Alcheringa' or 'Dream time' is to the Arunta. Asked for the reason why of anything, the Arunta answer, 'It was so in the Alcheringa.' Our tribe have a subsidiary myth corresponding to that of the Alcheringa. There was an age, in their opinion, when only birds and beasts were on earth; but a colossal man and two women came from the remote north-cast, changed birds and beasts into men and women, made other folk of clay or stone, taught them everything, and left laws for their guidance, then returned whence they came. This is a kind of 'Alcheringa' myth, but whether this colossal man was Byamee or not, our tribe give, as the final answer to any question about the origin of customs, 'Because Byamee say so.' Byamee declared his will, and that was and is enough for his children. At the Boorah, or initiatory ceremonies, he is proclaimed as 'Father of All, whose laws the tribes are now obeying.' Byamee, at least in one myth (told also by the Wirádjuri), is the original source of all totems, and of the law that people of the same totem may not intermarry, 'however far apart their hunting-grounds.' I heard first in a legend, then received confirmation from all old blacks, that Byamee had a totem name for every part of his body, even to a different one for each finger and toe. And when he was passing on to fresh fields, he gave each kinship of the tribe he was leaving one of his totems. The usual version is, that to such as were metamorphosed from birds and animals he gave as totem the animal or whatever it was from which they were evolved. But no one dreams of claiming Byamee as a relation belonging to one clan; he is one apart and yet the father of all, even as Birrahgnooloo is mother of all and not related to any one clan; Cunnumbeillee, his other wife, had only one totem.

Certainly woman is given a high place in their sacred lore. The chief wife of Byamee, Birrahgnooloo, is claimed as the mother of all, for she, like him, had a totem for each part of her body; no one totem can claim her, but all do.

Mother of all, though mother of none in particular, she was not to be vulgarised by ordinary domestic relations, For those purposes Cunnumbeillee was at hand, as a bearer of children and a caterer. Yet it was Birrahgnooloo whom Byamee best loved and made his companion, giving her power and position which no other held. She too, like him, is partially crystallised in the sky-camp, where they are together; the upper parts of their bodies are as on earth; to her, those who want floods go, and when willing to grant their requests, she bids Cunnumbeillee start the flood-ball of blood rolling down the mountains. Cunnumbeillee, as has been said, had but one totem which her children derived from her.

Byamee is the originator of things less archaic and important than totemism. There is a large stone fish-trap at Brewarrina, on the Barwan River. It is said to have been made by Byamee and his gigantic sons, just as later Greece attributed the walls of Tiryns to the Cyclops, or as Glasgow Cathedral has been explained in legend as the work of the Picts. Byamee also established the rule that there should be a common camping-ground for the various tribes, where, during the fishing festival, peace should be strictly kept, all meeting to enjoy the fish, and do their share towards preserving the fisheries.

Byamee still exists. I have been told by an old native, as will be shown later, that prayers for the souls of the dead used to be addressed to Byamee at funerals; certainly not a practice derived from Protestant missionaries.

Byamee is supposed to listen to the cry of an orphan for rain. Such an one has but to run out when the clouds are overhead, and, looking at the sky, call aloud

'Gullee boorboor. Gullee boorboor.'
'Water come down. Water come down.'

Or should it be raining too much, the last possible child of a woman can stop it by burning Midjeer wood.

Bootha told me after one rain that she had sent one of her tutelary spirits to tell Boyjerh--Byamee is called by women and children Boyjerh--that the country wanted rain. In answer he had taken up a handful of crystal pebbles and thrown them from the sky down into the water in a stone basin on the top of the sacred mountain; as the pebbles fell in, the water splashed up into the clouds above, whence it descended as the desired rain.

It is told to me, that at some initiatory rites the oldest medicine man, or Wirreenun, present addresses a prayer to Byamee, asking him to give them long life, as they have kept his law.

The tribesmen do not profess to pray, or to have prayed, to Byamee on any occasions except at funerals, and at the conclusion of the Boorah.

As for Byamee's relation to ethics, it will be stated in the chapter on the tribal ceremonies, while the stories as to the rewards and punishments of the future life will be given in their place. Baiame's troubles with a kind of disobedient deputy, Darramulun, will also be narrated: the myth is current, too, among the Wirádjuri tribe.

Other particulars about Byamee will occur in the course of later chapters: here I have tried to give a general summary of the native beliefs. The reader may interpret them in his own fashion, and may decide as to whether the beliefs do or do not indicate a kind of 'religion,' whether 'a recognised religion' or not. There is necessarily, of course, an absence of temples and of priests, and I have found no trace or vestige of sacrifice. What may be said on the affirmative side as to the religious aspect of the belief, the reader can supply from the summary of facts. Other potent beings occur in native myth, as we shall show, but there appears to exist between them and mankind no relation of affection, reverence, or duty, as in the case of Byamee.

Here it seems necessary to advert to a remark of Mr. Howitt's which appears to be erroneous. He says 'that part of Australia which I have indicated as the habitat of that belief' (namely, in an All Father),' is also the area where there has been the advance from group marriage to individual marriage; from descent in the female line to that in the male line; where the primitive organisation under the class system has been more or less replaced by an organisation based on locality; in fact, where these advances have been made to which I have more than once drawn attention.'[21]

Mr. Howitt forgets that he himself attributes the early system of descent through women, and also the belief in an All Father (Nurelli), to the Wiimbaio tribe[22] to the Wotjobaluk tribe,[23] to the Kamilaroi, to the Ta-Ta-thi,[24] while female descent and the belief in Baiame mark the Euahlayi and Wirádjuri.[25]

These tribes cover an enormous area of country, and, though they have not advanced to male kinship, they all possess the belief in an All Father. That belief does not appear to be in any way associated with advance in social organisation, for Messrs. Spencer and Gillen cannot find a trace of it in more than one of the central and northern tribes, which have male kinship, and a kind of local self-government. On the other hand, it does occur among southern tribes, like the Kurnai, which have advanced almost altogether out of totemism.

In short, we have tribes with female descent, such as the Dieri and Urabunna, to whom all knowledge of an All Father is denied. We have many large and important tribes with female descent who certainly believe in an All Father. We have tribes of the highest social advancement who are said to show no vestige of the belief, and we have tribes also socially advanced who hold the belief with great vigour. In these circumstances, authenticated by Mr. Howitt himself, it is impossible to accept the theory that belief in an All Father is only reached in the course of such advance to a higher social organisation as is made by tribes who reckon descent in the male line.

RELATIONSHIPS AND TOTEMS

SOME savants question the intellectual ability of the blacks because they have not elaborate systems of numeration and notation, which in their life were quite unneeded. Such as were needed were supplied. They are often incorporate in one word-noun and qualifying numerical adjective, as for example--

Gundooee	*a solitary emu*
Booloowah	*two emus*
Oogle oogle	*four emus*
Gayyahnai	*five or six emus*
Gonurrun	*fourteen or fifteen emus.*

I fancy the brains that could have elaborated their marriage rules were capable of workaday arithmetic if necessary, and few indeed of us know our family trees as the blacks know theirs.

Even the smallest black child who can talk seems full of knowledge as to all his relations, animate and inanimate, the marriage taboos, and the rest of their complicated system.

The first division among this tribe is a blood distinction (I phratries'):--

Gwaigulleeah	*light blooded*
Gwaimudthen	*dark blooded.*

This distinction is not confined to the human beings of the tribe, who must be of one or the other, but there are the Gwaigulleeah and Gwaimudthen divisions in all things. The first and chief division in our tribe, as regards customary marriage law, is the partition of all tribes-folk into these 'phratries,' or 'exogamous moieties.' While in most Australian tribes the meanings of the names of phratries are lost, where the meanings are known they are usually names of animals-Eagle, Hawk, and Crow, White Cockatoo and Black Cockatoo, and so forth. Among the great Kamilaroi tribe, akin in speech to the Euahlayi, the names of phratries, Dilbi and Kupathin, are of unknown significance. The Euahlayi names, we have seen, are Gwaigulleeah, Light blooded, and Gwaimudthen, Dark blooded.

The origin of this division is said to be the fact that the original ancestors were, on the one side, a red race coming from the west, the Gwaigulleeah; on the other, a dark race coming from the east.

A Gwaigulleeah may under no circumstances marry a Gwaigulleeah; he or she must mate with a Gwaimudthen. This rule has no exception. A child belongs to the same phratry as its mother.

The next name of connection is local, based on belonging to one country or hunting-ground; this name a child takes from its mother wherever it may happen to be born. Any one who is called a Noongahburrah belongs to the Noongah-Kurrajong country; Ghurreeburrah to the orchid country; Mirriehburrah, poligonum country; Bibbilah, Bibbil country, and so on. This division, not of blood relationship, carries no independent marriage restriction, but keeps up a feeling equivalent to Scotch, Irish, or English, and is counted by the blacks as 'relationship,' but not sufficiently so to bar marriage.

The next division is the name in common for all daughters, or all sons of one family of sisters. The daughters take the name from their maternal grandmother, the sons from their maternal great-uncle.

Of these divisions, called I Matrimonial Classes, there are four for each sex, bearing the same names as among the Kamilaroi. The names are--

Masculine	Kumbo	brother and sister
Feminine,	Bootha	
Masculine	Murree	brother and sister
Feminine,	Matha	
Masculine	Hippi	brother and sister
Feminine,	Hippitha	
Masculine	Kubbee	brother and sister
Feminine,	Kubbootha	

The children of Bootha will be

Masculine	Hippi	brother and sister
Feminine,	Hippitha	

The children of Matha will be

Masculine	Kubbee	brother and sister

Feminine,	Kubbootha

The children of Hippatha will be

Masculine	Kumbo	brother and sister
Feminine,	Bootha	

The children of Kubbootha will be

Masculine	Murree	brother and sister
Feminine,	Matha	

Thus, you see, they take, if girls, their grandmother's and her sisters' 'class' names in common; if boys, the 'class' name of their grandmother's brothers.

Bootha	Can only marry	Murree,
Matha	"	Kumbo,
Hippitha	"	Kubbee,
Kubbootha	"	Hippi.

Both men and women are often addressed by these names when spoken to.

A propos of names, a child is never called at night by the same name as in the daytime, lest the 'devils' hear it and entice him away.

Names are made for the newly born according to circumstances; a girl born under a Dheal tree, for example, was called
Dheala. Any incident happening at the time of birth may gain a
child a name, such as a particular lizard passing. Two of my
black maids were called after lizards in that way: Barahgurree
and Bogginbinnia.

Nimmaylee is a porcupine with the spines coming; such an one
having been brought to the camp just as a girl was born, she
became Nimmaylee.

The mothers, with native politeness, ask you to give their
children English names, but much mote often use in familiar
conversation either the Kumbo Bootha names, or others
derived from place of birth, from some circumstance connected
with it, a child's mispronunciation of a word, some peculiarity
noticed in the child, or still more often they call each other by
the name proclaiming the degree of relationship.

For example, a girl calls the daughters of her mother and of her
aunts alike sisters.

Boahdee	sister
Wambaneah	full brother
Dayadee	half brother
Gurrooghee	uncle
Wulgundee	uncle's wife
Kummean	sister's sister
Numbardee	mother

Numbardee	mother's sister
Beealahdee	Father
Beealahdee	Mother's sisters' husbands
Gnahgnahdee	Grandmother on father's side
Bargie	Grandmother on mother's side
Dadadee	Grandfather on mother's side
Gurroomi	a son-in-law, or one who could be a son-in-law
Goonooahdee	a daughter-in-law, or one who could be a daughter-in-law
Gooleerh	Husband or wife, or one who might be so.

So relationships are always kept in their memories by being daily used as names. There are other general names, too, such as--

Mullayerh	a temporary mate or companion
Moothie	a friend of childhood in after life
Doore-oothai	a lover
Dillahga	an elderly man of the same totem
Tuckandee	a young man of the same totem, reckoned as a sort of brother.

Another list of names used ordinarily is--

Boothan	last possible child of a woman
Mahmee	old woman
Beewun	motherless girl
Gowun	fatherless girl
Yumbui	fatherless boy
Moogul	only child.

Those of the same totem are reckoned as brothers and sisters, so cannot intermarry. 'Boyjerh' relations, as those on the father's side are called, are not so important as on the mother's side, but are still recognised.

Now for the great Dhé or totem system, by some called Mah, but Dhé is the more correct.

Dinewan, or emu, is a totem, and has amongst its multiplex totems' or 'sub-totems'--

Goodoo	or codfish
Gumbarl	silver bream
Inga	crayfish
Boomool	shrimps
Gowargay	water emu spirit
Moograbah	big black-and-white magpie

Booloorl	little night owl
Byahmul	black swan
Eerin	a little night owl
Beerwon	a bird like a swallow
Dulloorah	the manna-bringing birds
Bunnyal	flies
Dheal	sacred fire
Gidya	an acacia
Yaraan	an eucalyptus
Deenyi	ironbark
Guatha	quandong
Goodooroo	river box
Mirieh	poligonum
Yarragerh	the north-east wind
Guie	tree--Owenia acidula
Niune	wild melon
Binnamayah	big saltbush.

Bohrah, the kangaroo, is another totem, and is considered somewhat akin to Dinewan. For example, in a quarrel between, say, the Bohrah totem and the Beewee, the Dinewan would take the part of the former rather than the latter.

Amongst the multiplex totems of Bohrah are--

Goolahwilleel	topknot pigeons
Boogoodoogadah	the rain-bird
Gilah	fink-breasted parrot
Quarrian	yellow and red breasted grey parrot green
Buln Buln	parrot
Gidgerregah	small green parrot
Cocklerina	a rose and yellow crested white cockatoo
Youayah	frogs
Guiggahboorool	biggest ant-beds
Dunnia	wattle tree
Mulga	an acacia
Gnoel	sandalwood
Brigalow	an acacia
Yarragerh	north-east wind, same as Dinewan's.

All clouds, lightning, thunder, and rain that is not blown up by the wind of another totem, belong to Bohrah.

Beewee, brown and yellow Iguana, numerically a very powerful totem, has for multiplex totems--

Gai-gai	catfish
Curreequinquin	butcher-bird
Gougourgahgah	laughing-jackass
Deenbi	divers
Birroo Birroo	sand builders
Deegeenboyah	soldier-bird
Weedah	bower-bird
Mooregoo Mooregoo	black ibis
Booloon	white crane
Noodulnoodul	whistling ducks
Goborrai	stars
Gulghureer	pink lizard
Goori	pine
Talingerh	native fuchsia
Guiebet	native passion fruit
Boonburr	poison tree
Gungooday	stockman's wood

Guddeeboondoo	bitter bark
Boorgoolbean or Mooloowerh	a shrub with creamy blossoms
Yarragerh	spring wind
Muddernwurderh	west wind.

Those with whom the Beewee shares the winds he counts as relations. It is the Beewees of the Gwaimudthen, or dark blood, who own Yarragerh (spring wind); the light-blooded own Mudderwurderh (west wind).

Another totem is Gouyou, or Bandicoot. The animal has disappeared from the Narran district, but the totem tribe is still strong, though not so numerous as either the Beewees or Dinewans.

Multiplex totems of Gouyou--

Wayarnberh	turtle
Mungghee	mussels
Piggiebillah	Porcupine
Dayahminnah	small carpet snake
Mungun	large carpet snake
Douyouie	ants
Moondoo	wasps

Murgahmuggui	spider
Bayarh	green-head ants
Mubboo	beefwood
Coolabah	eucalyptus, flooded box
Bingahwingul	needlebush
Mayarnah	stones
Gheeger Gheeger	cold west wind
Gibbon	yam
Boondoon	kingfisher
Durnerh brown	pigeon
Guineeboo	redbreasts
Munggheewurraywurraymul	seagulls
Guiggah ordinary	ant-beds.

Next we take Doolungaiyah, or Bilber, commonly known as Bilby, a large species of rat the size of a small rabbit, like which it burrows; almost died out now. The totem clan are very few here too, so it is difficult to learn much as to their multiplex totems, amongst which, however, are--

Ooboon	blue-tongued lizard
Goomblegubbon	plains turkey or bustard

Boothagullagulla	bird like seagull
Tekel Barain	large white amaryllis.

Douyou, black snake, totem claims--

Noongah	kurrajong--sterculia
Carbeen	an eucalyptus
Booroorerh	bulrushes
Gargooloo	yams
Yhi	the sun (feminine)
Gunyahmoo	the east wind
Kurreah	crocodile
Wa-ah	shells
Douyougurrah	earth-worms
Deereeree	willy wagtail
Burrengeen	jeewee
Bouyoudoorunnillee	grey cranes
Ouyan	curlew
Bouyougah	centipedes
Bubburr	big snake

Woggoon	scrub turkey
Beeargah	crane
Waggestmul	kind of rat
Wi	small fish
Millan	small water-yam--sourtop

Moodai, or opossum, another totem, claims--

Bibbil	popular-leaved gum
Bumble	Capparis Mitchellianni
Birah	whitewood
Beebuyer	yellow flowering broom
Illay	hop bush
Mirrie	wild currant bush
Mooregoo	swamp oak--belah
Mungoongarlee	largest iguana
Mouyi	white cockatoo
Beeleer	black cockatoo
Wungghee	white night owl
Mooregoo	mopoke
Narahdarn	bat

Bahloo	moon
Euloowirrie	rainbow
Bibbee	woodpecker
Billai	crimson wing parrot
Durrahgeegin	green frog.

Maira, a paddy melon, claims as multiplex totems--

Wahn	the crow
Mullyan	the eagle-hawk
Gooboothoo	doves
Goolayyalilee	pelican
Oonaywah	black diver
Gunundar	white diver
Birriebungar	small diver
Mounin	mosquito
Mouninguggahgui	mosquito bird
Bullah Bullah	butterflies
Tucki	a kind of bream
Beewerh	bony bream

Gulbarlee	shingleback lizard
Budtha	rosewood
Goodoogah	yalli
Wayarah	wild grapes
Garwah	rivers
Gooroongoodilbaydilbay	south wind.

It is said a Maira will never be drowned, for the rivers are a sub-totem of theirs; but I notice they nevertheless learn to swim.

Yubbah, carpet snake, as a kin has almost disappeared, only a few members remaining to claim

Mungahran	hawk.

Burrahwahn, a big sandhill rat, now extinct here, claims--

Mien	dingo
Dalleerin	a lizard
Gaengaen	wild lime
Willerhderh, or Douran Douran	north wind
Bralgah	native companion.

Buckandee, native cat kin, claim--

Buggila	leopard wood
Bean	myall
Bunbundoolooey	a little brown bird
Dunnee Bunbun	a very large green parrot
Dooroongul	hairy caterpillar.

Amongst other totems were once the Bralgah, Native Companion, and Dibbee, a sort of sandpiper, but their kins are quite extinct as far as our blacks are concerned; the birds themselves are still plentiful. The Bralgah birds have a Boorah ground at the back of our old horse-paddock, a smooth, well-beaten circle, where they dance the grotesque dances peculiar to them, which are really most amusing to watch, somewhat like a set of kitchen lancers into which some dignified dames have got by mistake, and a curious mixture is the dance of dignity and romping.

The totem kins numerically strongest with us were the Dinewans, Beewees, Bohrahs, and Gouyous. Further back in the country, they tell me, the crow, the eaglehawk, and the bees were original totems, not multiplex ones, as with us.

It may be as well for those interested in the marriage law puzzles to state that Dinewans, Bohrahs, Douyous, and Doolungayers are always

| Kumbo | Hippi |
| Bootha | Hippitha. |

That Moodai, Gouyou, Beewee, Maira, Yubbah are always

Murree   Kubbee

Matha   Kubbootha.

Our blacks may and do eat their hereditary totems, if so
desirous, with no ill effects to themselves, either real or
imaginary; their totem names they take from their mothers.
They may, in fact, in any way use their totems, but never abuse
them. A Beewee, for example, may kill, or see another kill, and
eat or use a Beewee, or one of its multiplex totems, and show
no sign of sorrow or anger but should any one speak evil of the
Beewee, or of any of its multiplex totems, there will be a
quarrel.

There will likewise be a quarrel if any one dares to mimic a
totem, either by drawing one, except at Boorahs, or imitating it
in any way.

There are members of the tribes, principally wizards, or men
intended to be such, who are given an individual totem called
Yunbeai. This they must never eat or they will die. Any injury to
his yunbeai hurts the man himself In danger he has the power
to assume the shape of his yunbeai, which of course is a great
assistance to him, especially in legendary lore; but, on the other
hand, a yunbeai is almost a Heel of Achilles to a wirreenun (see
the chapter on Medicine and Magic).

Women are given a yunbeai too, sometimes. One girl had a
yunbeai given her as a child, and she was to be brought up as a
witch, but she caught rheumatic fever which left her with St.
Vitus's dance. The yunbeai during one of her bad attacks

jumped out of her, and she lost her chance of witchery. One old fellow told me once that when he was going to a public-house he took a miniature form of his yunbeai, which was the Kurrea--crocodile-out of himself and put it safety in a bottle of water, in case by any chance he got drunk, and an enemy, knowing his yunbeai, coaxed it away. I wanted to see that yunbeai in a bottle, but never succeeded.

The differences between the hereditary totem or Dhé, inherited from the mother, and the individual totem or yunbeai, acquired by chance, are these: Food restrictions do not affect the totem, but marriage restrictions do; the yunbeai has no marriage restrictions; a man having an opossum for yunbeai may marry a woman having the same either as her yunbeai or hereditary totem, other things being in order, but under no circumstances must a yunbeai be eaten by its possessor.

The yunbeai is a sort of alter ego; a man's spirit is in his yunbeai, and his yunbeai's spirit in him.

A Minggah, or spirit-haunted tree of an individual, usually chosen from amongst a man's multiplex totems, is another source of danger to him, as also a help.

As Mr. Canton says: 'What singular threads of superstition bind the ends of the earth together! In an old German story a pair of lovers about to part chose each a tree, and by the tree of the absent one was the one left to know of his wellbeing or the reverse. In time his tree died, and she, hearing no news of him, pined away, her tree withering with her, and both dying at the same time.

Well, that is just what a wirreenun would believe about his Minggah. These Minggah and Goomarh spirit trees and stones always make me think, perhaps irrelevantly, of one of the restored sayings of the Lord, which ends 'Raise the stone, and there thou shalt find Me; cleave the wood, and I am there.'

Blacks were early scientists in some of their ideas, being before Darwin with the evolution theory, only theirs was a kind of evolution aided by Byamee. I dare say, though, the missing link is somewhere in the legends. I rather think the Central Australians have the key to it. One old man here was quite an Ibsen with his ghastly version of heredity.

He said, when I asked him what harm it would do for, say, a Beewee totem man to come from the Gulf country, where his tribe had never had any communication with ours, and marry a girl here,--that all Beewees were originally changed from the Beewee form into human shape. The Beewee of the Gulf, originally, like the Beewee here, had the same animal shape, and should two of this same blood mate the offspring would throw back, as they say of horses, to the original strain, and partake of iguana (Beewee) attributes either in nature or form.

From the statements just given, it will be seen that the Euahlayi are in the Kamilaroi stage of social organisation. They reckon descent in the female line: they have 'phratries' and four matrimonial classes, with totems within the phratries. In their system of 'multiplex-totems' or 'sub-totems' they resemble the Wotjobaluk tribe.[26] The essence of the 'sub-totem' system is the division of all things into the categories provided by the social system of the human society. The arrangement is a very early attempt at a scientific system of classification.

Perhaps the most peculiar feature in the organisation of the Euahlayi is the existence of Matrimonial Classes, which are named as in the Kamilaroi tongue, while the phratry names are not those of the Kamilaroi, and alone among phratry names in

Australia which can be translated, are not names of animals. The phratries have thus no presiding animals, and in the phratries there are no totem kins of the phratriac names. The cause of these peculiarities is matter of conjecture.

A peculiarity in the totemic system of the Euahlayi--the right of each individual to kill and eat his own totem--has been mentioned, and may be associated here with other taboos on food.

The wunnarl, or food taboo, was taken off a different kind of food for boys at each Boorah, until at last they could eat what they pleased except their yunbeai, or individual familiar: their Dhé, or family totem, was never wunnarl or taboo to them.

A child may not perhaps know that it has had a yunbeai given to it, and may eat of it in ignorance, when immediately they say that child sickens.

Should a boy or a girl eat plains turkey or bustard eggs while they were yet wunnarl, or taboo, he or she would lose his or her sight. Should they eat the eggs or flesh of kangaroo or piggiebillah, their skins would break out in sores and their limbs wither.

Even honey is wunnarl at times to all but the very old or very young. Fish is wunnarl for about four years after his Boorah to a boy, and about four months after she is wirreebeeun, or young woman, to a girl.

When the wunnarl was taken off a particular kind of meat, a wizard poured some of the melted fat and inside blood of that animal or bird, as the case might be, over the boy, and rubbed it into him. The boy, shaking and shivering, made a spluttering noise with his lips; after that he could eat of the hitherto forbidden food.

This did not necessarily refer to his totem, but any food wunnarl to him, though it is possible that there may have been a time in tribal history, now forgotten, when totems were wunnarl, and these ceremonies may be all that is left to point to that time.

When a boy, after his first Boorah, killed his first emu, whether it was his Dhé, or totem, or not, his father made him lie on the bird before it was cooked. Afterwards a wirreenun (wizard) and the father rubbed the fat on the boy's joints, and put apiece of the flesh in his mouth. 'The boy chewed it, making a noise as he did so of fright and disgust; finally he dropped the meat from his mouth, making a blowing noise through his lips of 'Ooh! Ooh! Ooh!' After that he could eat the flesh.

A girl, too, had to be rubbed with the fat and blood of anything from which the wunnarl was to be removed for her. No ceremony of this sort would be gone through with the flesh, fat, or blood of any one's yunbeai, or individual familiar animal, for under no circumstances would any one kill or eat their yunbeai.

Concerning the yunbeai, or animal familiar of the individual, conferred by the medicine men, more is to be said in the ensuing chapters. The yunbeai answers to the Manitu obtained by Red Indians during the fast at puberty; to the 'Bush Soul' of West Africa; to the Nagual of South American tribes; and to the Nyarong of Borneo. The yunbeai has hitherto been scarcely remarked on among Australian tribes. Mr. Thomas declares it to be 'almost non-existent' in Australia, mentioning as exceptions its presence among the Euahlayi; the Wotjobaluk in Victoria; the Yaraikkanna of Cape York; and 'probably' some of the northern tribes on the other side of the Gulf of Carpentaria.[27]

Perhaps attention has not been directed to the animal familiar in Australia, or perhaps it is really an infrequent thing among the tribes.

THE MEDICINE MEN

I USED to wonder how the wirreenuns or doctor-wizards of the tribe attained their degrees.

I found out that the old wizards fix upon a young boy who is to follow their profession. They take him to a tribal burial-ground at night. There they tie him down and leave him , after having lit some fires of fat at short distances round him.

During the night that boy, if he be shaky in his nerves, has rather a bad time.

One doctor of our tribe gave me a recital of his own early experience.

He said, after the old fellows had gone, a spirit came to him, and without undoing his fastenings by which he was bound, turned him over, then went away. Scarcely had the spirit departed when a big star fell straight from the sky alongside the boy; he gazed fixedly at it, and saw emerge from it, first the two hind legs, then the whole of a Beewee or iguana. The boy's totem was a Beewee, so he knew it would not hurt him. It ran close up to him, climbed on him, ran down his whole length, then went away.

Next came a snake straight towards his nose, hissing all the time. He was frightened now, for the snake is the hereditary enemy of the iguana. The boy struggled to free himself, but ineffectually. He tried to call out but found himself dumb. He tried to shut his eyes, or turn them from the snake, but was powerless to do so. The snake crawled on to him and licked him. Then it went away, leaving the boy as one paralysed. Next came a huge figure to him, having in its hand a gunnai or yam stick. The figure drove this into the boy's head, pulled it out through his back, and in the hole thus made placed a 'Gubberah,' or sacred stone, with the help of which much of the boy's magic in the future was to be worked.

This stone was about the size and something the shape of a small lemon, looking like a smoothed lump of semi-transparent crystal. It is in such stones that the wi-wirreenuns, or cleverest wizards, see visions of the past, of what is happening in the present at a distance, and of the future; also by directing rays from them towards their victims they are said to cause instantaneous death.

Next, to the doctor-boy on trial, came the spirits of the dead who corroboreed round him, chanting songs full of sacred lore as regards the art of healing, and instructions how, when he needed it, he could call upon their aid.

Then they silently and mysteriously disappeared. The next day one of the old wizards came to release the boy; he kept him away from the camp all day and at night took him to a weedah, or bower-bird's, playground. There he tied him down again, and there the boy was visited again by the spirits of the dead, and more lore was imparted to him.

The reason given for taking him to a weedah's playground is, that before the weedah was changed into a bird, he was a great wirreenun; that is why, as a bird, he makes such a collection of pebbles and bones at his playground.

The bower-bird's playgrounds are numerous in the bush. They are made of grass built into a tent-shaped arch open at each end, through which the weedahs run in and out, and scattered in heaps all around are white bones and black stones, bits of glass, and sometimes we have found coins, rings, and brooches.

The weedahs do not lay their eggs at their playgrounds their nests are hard to find. A little boy always known as 'Weedah,' died lately, so probably a new name will have to be found for the bird, or to mention it will be taboo, at all events before the old people, who never allow the names of the dead to be mentioned.

For several nights the medical student was tied down in case he should be frightened and run away, after that he was left without bonds. He was kept away from the camp for about two months. But he was not allowed to become a practitioner until he was some years older: first he dealt in conjuring, later on he was permitted to show his knowledge of pharmacy.

His conjuring cures are divers.

A burn he cures by sucking lumps of charcoal from it. Obstinate pains in the chest, the wizard says, must be caused by some enemy having put a dead person's hair', or bone in it. Looking wisdom personified in truly professional manner, he sucks at the affected spot, and soon produces from his mouth hair, bones, or whatever he said was there.

If this faith-healing does not succeed, a stronger wizard than he must have bewitched the patient; he will consult the spirits. To that end he goes to his Minggah, a tree or stone-more often a tree, only the very greatest wirreenuns have stones, which are called Goomah--where his own and any spirits friendly towards him may dwell.

He finds out there who the enemy is, and whence he obtained his poison. If a wirreenun is too far away to consult his friendly spirits in person, he can send his Mullee Mullee, or dream spirit, to interview them.

He may learn that an enemy has captured the sick person's Doowee, or dream spirit--only wirreenuns' dream spirits are Mullee Mullee, the others are Doowee--then he makes it his business to get that Doowee back.

These dream spirits are rather troublesome possessions while their human habitations sleep they can leave them and wander at will. The things seen in dreams are supposed to be what the Doowees see while away from the sleeping bodies. This wandering of the Doowees is a great chance for their enemies: capture the Doowee and the body sickens; knock the Doowee about before it returns and the body wakes up tired and languid. Should the Doowee not return at all, the person from whom it wandered dies. When you wake up unaccountably tired in the morning, be sure your Doowee has been 'on the spree,' having a free fight or something of that sort. And though your Doowee may give you at times lovely visions of passing paradises, on the whole you would be better without him.

There is on the Queensland border country a dillee bag full of unclaimed Doowees. The wirreenun who has charge of this is one of the most feared of wirreenuns; he is a great magician, who, with his wonder-working glassy stones, can conjure up visions of the old fleshly habitations of the captured Doowees.

He has Gubberahs, or clever stones, in which are the active spirits of evil-working devils, as well as others to work good. Should a Doowee once get into this wirreenun's bag, which has the power of self-movement, there is not a great chance of getting it back, though it is sometimes said to be done by a rival combination of magic. The worst of it is that ordinary people have no power over their Doowees; all they can do is to guard against their escaping by trying to keep their mouths shut while asleep.

The wirreenuns are masters of their Mullee Mullees, sending them where they please, to do what they are ordered, always provided they do not meet a greater than themselves.

All sorts of complications arise through the substitution of mad or evil spirits for the rightful Doowee. Be sure if you think any one has suddenly changed his character unaccountably, there has been some hankey-pankey with that person's Doowee. One of the greatest warnings of coming evil is to see your totem in a dream; such a sign is a herald of misfortune to you or one of your immediate kin. Should a wirreenun, perhaps for enmity, perhaps for the sake of ransom, decide to capture a Doowee, he will send his Mullee Mullee out to do it, bidding the Mullee Mullee secrete the Doowee in his--the wirreenun's--Minggah, tree or rock.

When he is consulted as to the return of the missing Doowee, he will order the one who has lost it to Sleep, then the Doowee, should the terms made suit the wirreenun, re-enters the body. Should it not do so, the Doowee-less one is doomed to die.

In a wirreenun's Minggah, too, are often secreted shadow spirits stolen from their owners, who are by their loss dying a lingering death, for no man can live without Mulloowil, his shadow. Every one has a shadow spirit which he is very careful not to parade before his enemies, as any injury to it affects himself. A wirreenun can gradually shrink the shadow's size, the owner sickens and dies. 'May your shadow never be less!'

The shadow of a wirreenun is, like his head, always mahgarl, or taboo; any one touching either will be made to suffer for such sacrilege.

A man's Minggah is generally a tree from amongst his multiplex totems,' as having greater reason to help him, being of the same family.

In his Minggah a wirreenun will probably keep some Wundah, or white devil spirits, with which to work evil. There, too, he often keeps his yunbeai, or animal spirit--that is, his individual totem, not hereditary one. All wirreenuns have a yunbeai, and sometimes a special favourite of the wirreenuns is given a yunbeai too--or in the event of any one being very ill, he is given a yunbeai, and the strength of that animal goes into the patient, making him strong again, or a dying wirreenun leaves his yunbeai to some one else. Though this spirit gives extra strength it likewise gives an extra danger, for any injury to the animal hurts the man too; thus even wirreenuns are exposed to danger.

No one, as we have said, must eat the flesh of his yunbeai animal; he may of his family totem, inherited from his mother, but of his yunbeai or individual familiar, never.

A wirreenun can assume the shape of his yunbeai; so if his yunbeai were, for example, a bird, and the wirreenun were in danger of being wounded or killed, he would change himself into that bird and fly away.

A great wirreenun can substitute one yunbeai for another, as was done when the opossum disappeared from our district, and the wirreenun, whose yunbeai it was, sickened and lay ill for months. Two very powerful wirreenuns gave him a new yunbeai, piggiebillah, the porcupine. His recovery began at once. The porcupine had been one of his favourite foods; from the time its spirit was put into him as his yunbeai, he never touched it.

A wirreenun has the power to conjure up a vision of his particular yunbeai, which he can make visible to those whom he chooses shall see it.

The blacks always told me that a very old man on the Narran, dead some years ago, would show me his yunbeai if I wished; it was Oolah, the prickly lizard.

One day I went to the camp, saw the old man in his usual airy costume, only assumed as I came in sight, a tailless shirt. One of the gins said something to him; he growled an answer; she seemed persuading him to do something. Presently he moved away to a quite clear spot on the other side of the fire; he muttered something in a sing-song voice, and suddenly I saw him beating his head as if in accompaniment to his song, and then--where it came from I can't say--there beside him was a lizard. That fragment of a shirt was too transparent to have hidden that lizard; he could not have had it up his sleeve, because his sleeves were in shreds. It may have been a pet lizard that he charmed in from the bush by his song, but I did not see it arrive.

They told me this old man had two yunbeai, the other was a snake. He often had them in evidence at his camp, and when he died they were seen beside him; there they remained until he was put into his coffin, then they disappeared and were never seen again. This man was the greatest of our local wizards, and I think really the last of the very clever ones. They say he was an old grey-headed man when Sir Thomas Mitchell first explored the Narran district in 1845. We always considered him a centenarian.

It was through him that I heard some of the best of the old legends, with an interpreter to make good our respective deficiencies in each other's language.

In the lives of blacks, or rather in their deaths, the Gooweera, or poison sticks or bones, play a great part.

A Gooweera is a stick about six inches long and half an inch through, pointed at both ends. This is used for sickening' or killing men.

A Guddeegooree is a similar stick, but much smaller, about three inches in length, and is used against women.

A man wishing to injure another takes one of these sticks, and warms it at a small fire he has made; he sticks the gooweera in the ground a few inches from the fire. While it is warming, he chants an incantation, telling who he wants to kill, why he wants to kill him, how long he wants the process to last, whether it is to be sudden death or a lingering sickness.

The chant over, and the gooweera warmed, he takes it from the fire. Should he wish to kill his enemy quickly, he binds opossum hair cord round the stick, only leaving one point exposed; should he only want to make his enemy ill, he only partially binds the stick. Then he ties a ligature tightly round his right arm, between the wrist and elbow, and taking the gooweera, or guddeegooree, according to the sex of his enemy, he points it at the person he wishes to injure, taking care he is not seen doing it.

Suddenly he feels the stick becoming heavier, he knows then it is drawing the blood from his enemy. The poison is prevented from entering himself by the ligature he has put round his arm. When the gooweera is heavy enough he ceases pointing it.

If he wants to kill the person outright, he goes away, makes a small hole in the earth, makes a fire beside it. In this hole he puts a few Dheal leaves--Dheal is the tree sacred to the dead; on top of the leaves he puts the gooweera, then more leaves this done, he goes away. The next day he comes back with his hand he hits the earth beside the buried stick, out jumps the gooweera, his enemy is dead. He takes the stick, which may be used many times, and goes on his way satisfied. Should he only wish to inflict a lingering illness on his enemy, he refrains from burying the gooweera, and in this case it is possible to save the afflicted person.

For instance, should any one suspect the man with the gooweera of having caused the illness, knowing of some grudge he had against the sick person, the one who suspects will probably intercede for mercy. The man may deny that he knows anything about it. He may, on the other hand, confess that he is the agent. If the intercessions prevail, he produces the gooweera, rubs it all over with iguana fat, and gives the intercessor what fat is left to rub over the sick person, who, on that being done, gradually regains his normal condition after having probably been reduced to a living skeleton from an indescribable wasting sickness, which I suspect we spell funk.

The best way to make a gooweera effective is to tie on the end of it some hair from the victim's head-a lock of hair being, in this country of upside-downs, a hate token instead of one of love.

When the lock of hair method is chosen as a means of happy dispatch, the process is carried out by a professional.

The hair is taken to the Boogahroo--a bag of hair and goowee-ras--which is kept by one or two powerful wirreenuns in a certain Minggah. The wirreenun on receiving the hair asks to whom it belongs. Should it belong to one of a tribe he is favourably disposed towards, he takes the gooweera or hair, puts it in the bag, but never sings the I death song' over it, nor does he warm it.

Should he, however, be indifferent, or ill-disposed towards the individual or his tribe, he completes the process by going through the form already given, or rather when there are two wirreenuns at the Boogahroo, the receiver of the hair gives it to the other one, who sings the death-song, warms the gooweera, and burns the hair. The person from whose head the hair on the gooweera came, then by sympathetic magic, at whatever distance he is, dies a sudden or lingering death according to the incantation sung over the poison-stick. Gooweeras need not necessarily be of wood; bone is sometimes used, and in these latter days even iron.

Sometimes at a large meeting of the blacks the Boogahroo wirreenuns bring the bag and produce from it various locks of hair, which the owners or their relations recognise, claim, and recover. They find out, from the wirreenun, who put them there; on gaining which knowledge a tribal feud is declared-a regular vendetta, which lasts from generation to generation.

If it be known that a man has stolen a lock of hair, he will be watched and prevented from reaching the Boogahroo tree, if possible.

These gooweeras used to be a terrible 'nuisance to us on the station. A really good working black boy would say he must leave, he was going to die. On inquiry we would extract the information that some one was pointing a gooweera at him.

Then sometimes the whole camp was upset; a strange black fellow had arrived, and was said to have brought gooweeras. This reaching the boss's ears, confiscation would result in order to restore peace of mind in the camp. Before I left the station a gin brought me a gooweera and told me to keep it; she had stolen it from her husband, who had threatened to point it at her for talking to another man.

Some of them, though they still had faith in the power of such charms, had faith also in me. I used to drive devils out with patent medicines; my tobacco and patent medicine accounts while collecting folk-lore were enormous.

A wirreenun, or, in fact, any one having a yunbeai, has the power to cure any one suffering an injury from whatever that yunbeai is; as, for example, a man whose yunbeai is a black snake can cure a man who is bitten by a black snake, the method being to chant an incantation which makes the yunbeai enter the stricken body and drive out the poison. These various incantations are a large part of the wirreenun's education; not least valuable amongst them is the chant sung over the tracks of snakes, which renders the bites of those snakes innocuous.

MORE ABOUT THE MEDICINE MEN AND LEECHCRAFT

THE wirreenuns sometimes hold meetings which they allow non-professionals to attend. At these the spirits of the dead speak through the medium of those they liked best on earth, and whose bodies their spirits now animate. These spirits are known as Yowee, the equivalent of our soul, which never leave the body of the living, growing as it grows, and when it dies take judgment for it, and can at will assume its perishable shape unless reincarnated in another form. So you see each person has at least three spirits, and some four, as follows: his Yowee, soul equivalent; his Doowee, a dream spirit; his Mulloowil, a shadow spirit; and may be his Yunbeai, or animal spirit.

Sometimes one person is so good a medium as to have the spirits of almost any one amongst the dead people speak through him or her, in the whistling spirit voice.

I think it is very clever of these mediums to have decided that spirits all have one sort of voice.

At these meetings there would be great rivalry among the wirreenuns. The one who could produce the most magical stones would be supposed to be the most powerful. The strength of the stones in them, whether swallowed or rubbed in through their heads, adds its strength to theirs, for these stones are living spirits, as it were, breathing and growing in their fleshly cases, the owner having the power to produce them at any time. The manifestation of such power is sometimes, at one of these trials of magic, a small shower of pebbles as seeming to fall from the heads and mouths of the rivals, and should by chance any one steal any of these as they fall, the power of the original possessor would be lessened. The dying bequeath these stones, their most precious possessions, to the living wirreenun most nearly related to them.

The wirreenun's health and power not only depend upon his crystals and yunbeai, but also on his Minggah; should an accident happen to that, unless he has another, he will die-in any case, he will sicken. Many of the legends deal with the magic of these spirit-animated trees.

They are places of refuge in time of danger; no one save the wirreenun, whose spirit-tree it was, would dare to touch a refugee at a Minggah; and should the sanctuary be a Goomarh, or spirit-stone, not even a wirreenun would dare to interfere, so that it is a perfectly safe sanctuary from humanly dealt evil. But a refugee at a Minggah or Goomarh runs a great risk of incurring the wrath of the spirits, for Minggah are taboo to all but their own wirreenun.

There was a Minggah, a great gaunt Coolabah, near our river garden. Some gilahs build in it every year, but nothing would induce the most avaricious of black bird-collectors to get the young ones from there.

A wirreenun's boondoorr, or dillee bag, holds a queer collection: several sizes of gooweeras, of both bone and wood, poison-stones, bones, gubberahs (sacred stones), perhaps a dillee--the biggest, most magical stone used for crystal-gazing, the spirit out of which is said to go to the person of whom you want to hear, wherever he is, to see what he is doing, and then show you the person in the crystal. A dinahgurrerhlowah, or moolee, death-dealing stone, which is said to knock a person insensible, or strike him dead as lightning would by an instantaneous flash.

To these are added in this miscellaneous collection medicinal herbs, nose-bones to put through the cartilage of his nose when going to a strange camp, so that he will not smell strangers easily. The blacks say the smell of white people makes them sick; we in our arrogance had thought it the other way on.

Swansdown, shells, and woven strands of opossum's hair are valuable, and guarded as such in the boondoorr, which is sometimes kept for safety in the wirreenun's Minggah.

Having dealt with the supernatural part of a wirreenun's training, which argues cunning in him and credulity in others, I must get to his more natural remedies.

Snakebite they cure by sucking the wound and cauterising it with a firestick. They say they suck out the young snakes which have been injected into the bitten person.

For headaches or pains which do not yield to the vegetable medicine, the wirreenuns tie a piece of opossum's hair string round the sore place, take one end in their mouths, and pull it round and round until it draws blood along the cord. For rheumatic pains in the head or in the small of the back and loins they often bind the places affected with coils of opossum hair cord, as people do sometimes with red knitting-silk.

The blacks have many herbal medicines, infusions of various barks, which they drink or wash themselves with, as the case may be.

Various leaves they grind on their dayoorl-stones, rubbing themselves with the pulp. Steam baths they make of pennyroyal, eucalyptus, pine, and others.

The bleeding of wounds they stanch with the down of birds.

For irritations of the skin they heat dwarf saltbush twigs and put the hot ends on the irritable parts.

After setting a broken limb they put grass and bark round it, then bind it up.

For swollen eyes they warm the leaves of certain trees and hold them to the affected parts, or make an infusion of Budtha leaves and bathe the eyes in it.

For rheumatic pains a fire is made, Budtha twigs laid on it, a little water thrown on them; the ashes raked out, a little more water thrown on, then the patient lies on top, his opossum rug spread over him, and thus his body is steamed. To induce perspiration, earth or sand is also often heated and placed in a hollowed-out space; on it the patient lies, and is covered with more heated earth.

Pennyroyal infused they consider a great blood purifier they also use a heap as a pillow if suffering from insomnia. It is hard to believe a black ever does suffer from insomnia, yet the cure argues the fact.

Beefwood gum is supposed to strengthen children. It is also used for reducing swollen joints. A hole is made in the ground, some coals put in, on them some beefwood leaves, on top of them the gum; over the hole is put enough bark to cover it with a piece cut out of it the size of the swollen joint to be steamed, which joint is held over this hole.

Various fats are also used as cures. Iguana fat for pains in the head and stiffness anywhere. Porcupine and opossum fats for preserving their hair, fish fat to gloss their skins, emu fat in cold weather to save their skins from chapping.

But what is supposed to strengthen them more than anything, both mentally and physically, is a small piece of the flesh of a dead person, or before a body is put in a bark coffin a few incisions were made in it; when it was coffined it was stood on end, and what drained from the incisions was caught in small wirrees and drunk by the mourners.

I fancy such cannibalism as has been in these tribes was not with a view to satisfaction of appetite but to the incorporation of additional strength. Either men or women are allowed to assist in this particularly nauseating funeral rite, but not the young people.

Nor must their shadows fall across any one who has partaken of this rite; should they do so some evil will befall them.

If the mother of a young child has not enough milk for its sustenance, she is steamed over 'old man' saltbush, and hot twigs of it laid on her breasts. To expedite the expulsion of the afterbirth, an old woman presses the patient round the waist, gives her frequent drinks of cold water, and sprinkles water over her. As soon as the afterbirth is removed a steam is prepared. Two logs are laid horizontally, some stones put in between them, then some fire, on top leaves of eucalyptus, and water is then sprinkled over them. The patient stands astride these logs, an opossum rug all over her, until she is well steamed. After this she is able to walk about as if nothing unusual had happened. Every night for about a month she has to lie on a steam bed made of damped eucalyptus leaves. She is not allowed to return to the general camp for about three months after the birth of her child.

Though perfectly well, she is considered unclean, and not allowed to touch anything belonging to any one. Her food is brought to her by some old woman. Were she to touch the food or food utensils of another they would be considered unclean and unfit for use. Her camp is gailie--that is, only for her; and she is goorerwon as soon as her child is born-a woman unclean and apart. Immediately a baby is born it is washed in cold water.

Ghastly traditions the blacks have of the time when Dunnerh-Dunnerh, the smallpox, decimated their ancestors. Enemies sent it in the winds, which hung it on the trees, over the camps, whence it dropped on to its victims. So terror-stricken were the tribes that, with few exceptions, they did not stay to bury their dead; and because they did not do so, flying even from the dying, a curse was laid on them that some day the plague would return, brought back by the Wundah or white devils; and the blacks shudder still, though it was generations before them, at the thought that such a horror may come again.

Poison-stones are ground up finely and placed in the food of the person desired to be got rid of. These poison-stones are of two kinds, a yellowish-looking stone and a black one; they cause a lingering death. The small bones of the wrist of a dead person are also pounded up and put into food, in honey or water, as a poison.

One cure struck me as quaint. The patient may be lying down, when up will come one of the tribe, most likely a wirreenun with a big piece of bark. He strikes the ground with this all round the patient, making a great row; this is to frighten the sickness away.

What seems to me a somewhat peculiar ceremony is the reception a coming baby holds before its birth.

The baby is presumably about to be born. Its grandmother is there naturally, but the black baby declines to appear at the request of its grandmother, and, moreover, declines to come if even the voice of its grandmother is heard; so grannie has to be a silent spectator while some other woman tempts the baby into the world by descanting on the glories of it. First, perhaps, she will say:

'Come now, here's your auntie waiting to see you.'

'Here's your sister.'

'Here's your father's sister,' and so on through a whole list. Then she will say, as the relatives and friends do not seem a draw:

'Make haste, the bumble fruit is ripe. The guiebet flowers are blooming. The grass is waving high. The birds are all talking. And it is a beautiful place, hurry up and see for yourself.'

But it generally happens that the baby is too cute to be tempted, and an old woman has to produce what she calls a wi-mouyan-a clever stick-which she waves over the expectant mother, crooning a charm which brings forth the baby.

If any one nurses a patient and the patient dies, the nurse wears an armlet of opossum's hair called goomil, and a sort of fur boa called gurroo.

If blacks go visiting, when they leave they make a smoke fire and smoke themselves, so that they may not carry home any disease.

As a rule blacks do not have small feet, but their hands are almost invariably small and well shaped, having tapering fingers.

OUR WITCH WOMAN

OUR witch woman was rather a remarkable old person. When she was, I suppose, considerably over sixty, her favourite granddaughter died.

Old Bootha was in a terrible state of grief, and chopped herself in a most merciless manner at the burial, especially about the head. She would speak to no one, used to spend her time about the grave, round which she fixed upright posts which she painted white, red, and black. All round the grave she used to sweep continually.

More and more she isolated herself, and at last discarded all her clothes and roamed the bush à la Eve before the Fall, as she had probably done as a young girl.

She dug herself an underground camp, roofed it over, and painted enormous posts which she erected in front of her 'Muddy wine,' as she called her camp. She never came near the house, though we had been great friends before.

She used to prowl round the outhouses and pick up all sorts of things, rubbish for the most part, but often good utensils too; all used to be secreted in the underground camp. She never talked to any one, but used to mutter continually to herself and her dogs in an unknown tongue which only her dogs seemed to understand.

We thought she was quite mad.

One day, while we were playing tennis, she suddenly, muttering her strange language and dancing new corroboree steps, clad only in her black skin, came up. Matah told her to go away, but she only corroboreed round him and said she wanted to see me. I have the most morbid horror of lunacy in any form. I was once induced to go over a lunatic asylum--the horror of it haunts me still. However, I thought it would never do to show the coward I was, so though I felt as if I had been scooped out and filled up with ice, I went to her. She danced round me for a little time, then sidled up to me and said:

'Wahl you frightened, wahl me hurt you. I only womba--mad--all yowee--spirits--in me tell me gubbah--good-I lib 'long a youee; bimeby I come back big feller wirreenun; wahl you frightened? I not hurt you.'

And after crooning an accompaniment to her steps off she went, a strange enough figure, dancing and crooning as she went towards her camp; and not until the spirits gave up possession of her did she come near the house again.

One day she gave us a start. We were schooling a new team of four horses. The off-side leader had only been in once before, and was a brumby (horse run in from a wild mob). We had to pass Bootha's camp. I looked about as we neared it but saw nothing of her. Suddenly from the ground, as it seemed, out dashed the weird old figure, arms full of things, jabbering away at a great rate. Whiz came a tin plate past the leaders' heads; the offside horse reared and plunged and took some holding. Whiz came an old bill; then, one after another, a regular fusilade of various utensils.

It did not take us long to get past, but for as long as we could see the attack was kept up. Coming back we saw nothing of Bootha, and all the utensils had been picked up.

I used to tell the other blacks to see that Bootha had plenty of food. They said she was all right, the spirits were looking after her. Lunatics, from their point of view, are only persons spirit-possessed.

Gradually old Bootha, clothed as usual, came back about the place.

Strange stories came through the house blacks to me of old Bootha. She was very ill for a long time, then suddenly she recovered; not only recovered but seemed rejuvenated. We heard of wonderful cures she made; how she always consulted the spirits about any illness; how there were said to be spirits in some of her dogs; how she was now a rainmaker and, in fact, a fully fledged witch.

I was curious to see some of these wonders, so used to get the old woman to come up when any one was ill, consult her, and generally make much of her. There is no doubt she could diagnose a case well enough. Matah suffered a good deal with a constant pain in one knee, he was quite lame from it. He showed it to Bootha one day. She sang a song to her spirits, then said:

'Too muchee water there; you steam him, put him on hot rag; you drink plenty cold water, all lite dat go.'

As it happened a medical man was passing a few days afterwards with an insurance agent. Matah consulted him.

'Hum! Yes, yes. Hot fomentations to the place affected, poultices, a cooling draught. There's a stoppage of fluid at the knee-joint which must be dispersed.'

I thought Bootha ought to have been called in consultation.

A girl I had staying with me was taken suddenly and, to us, unaccountably ill. She was just able to get out of her room into the drawing-room, where she would lie back on the cushions of a lounge looking dreadfully limp and utterly washed out. Hearing of her illness old Bootha came up. I thought it might amuse Adelaide to see an old witch; she agreed, so I brought her in.

Bootha went straight up to the sick girl, expressed a few sympathetic sentences, then she said she would ask the spirits what had made Adelaide ill and what would cure her.

She moved my furniture until she left the centre of the room clear; she squatted down, and hanging her head began muttering in an unintelligible dialect. Presently her voice ceased and we heard from beside her a most peculiar whistling sort of voice, to which she responded, evidently interrogating. Again the whistling voice from further away. Bootha then told me she had asked a dead black fellow, Big Joe, to tell her what she wanted to know; but he could not, so now she was going to ask her dead granddaughter. Again she said a sort of incantation, and again, after a while, came the whistling voice reply-this time from another direction, not quite so loud. The same sort of thing was gone through with the same result.

Then Bootha said she would ask Guadgee, a black girl who had been one of my first favourites in the camp, and who had died a few years previously.

The whistling voice came from a third direction, though all the time I could see Bootha's lips moving.

Guadgee answered all she was asked. She said Adelaide was made ill because she had offended the spirits by bathing in the creek under the shade of a Minggah, or spirit-tree, a place tabooed to all but wirreenuns, or such as hold communion with spirits.

Of course, according to the blacks, to disturb a shadow is to hurt the original.

In this Minggah, Guadgee. said, were swarms of bees invisible to all but wirreenuns, and they are ready always to resent any insult to the Minggah or its shadow. These spirit-bees had entered Adelaide and secreted some wax on her liver; their bites, Guadgee said, were on her back.

Well, that can't be it, I said, I for you never did bathe in the shade of a Minggah; for, going as you always do with the house-girls, you are bound to be kept from such sacrilege; they would never dare such desecration.'

'Which is their Minggah? Is it a big Coolabah between the Bend and the garden?'

'Yes.'

'Then I did bathe there the last time I went down. I was up too late to go with the Black-but-Comelys, and as the sun was hot I went further round the point and bathed in the shade. And the bee-bites must be those horribly irritating pimples I have across my back.'

The cause of illness settled to her satisfaction, Bootha asked how to cure it. The patient was to drink nothing hot nor heating but as much cold water as she liked, especially a long drink before going to bed. Guadgee said she would come in the night when the patient was asleep and take the wax from her liver; she would sleep well and wake better in the morning.

Bootha got up then, came over to the patient, took her hand, rubbed it round the wrist several times, muttering an incantation; then saying she would see her again next day, off she went, taking, she told us, all the spirits away inside her, whence at desire they could be returned to such Minggah in their own Noorunbah, or hereditary hunting-grounds, as wirreenuns had placed them in, or to roam at their pleasure when not required by those in authority over spirits. Our old spiritualist denies us freedom even in the after-life she promises us.

Adelaide slept that night, looked a better colour the next morning, and rapidly recovered.

We think old Bootha must be a good physician and a ventriloquist, only I believe it is said ventriloquists cannot live long, and Bootha is now over eighty.

Others besides wirreenuns see spirits sometimes, but rarely, though wirreenuns are said to have the power to conjure them up in a form visible to ordinary eyes.

Babies are said to see spirits when they are smiling or crowing as if to themselves; it's to some spirit visible to them but to no one else.

When a baby opens his hands and shuts them again quickly, smiling all the while, that baby is with the spirits catching crabs!

Dogs see spirits; when they bark and howl suddenly and you see nothing about, it is because they have seen a spirit.

One person may embody many spirits, but such an one must be careful not to drink anything hot or heating, such would drive out the spirits at once. The spirits would never enter a person defiled by the white man's 'grog.'

Old Bootha had an interview with a very powerful spirit after she was ill, who told her that the spirit of her father was now in Bahloo, the moon; and that it was this spirit which had cured her, and if she kept his commands she would live for ever. The commands were never to drink 'grog,' never to wear red, never to eat fish. This was told her fifteen years ago, never once has she transgressed; her vigour for an old woman considerably over eighty is marvellous.

She was going away for a trip. Before going she said, as she would not be able to know when I wanted rain for my garden, she would put two posts in it which had in them the spirits of Kurreahs, or crocodiles. As these spirits required water I might be certain my tanks would never go dry while they were on guard. She asked one of my Black-but-Comelys, a very stalwart young woman, to help her lift one of these posts into the garden where she wanted to erect it. The girl took hold of one end, but in a little while dropped it, said it was too heavy. Old Bootha got furious.

'I get the spirits to help me,' she said, and started a little sing-song, then shouldered the post herself and carried it in. These posts are painted red, black, and white, with a snaky pattern, the Kurreah sign, on them. She also planted in my garden two other witch-poles, one painted red and having a cross-bar about midway down it from which raddled strings were attached to the top; this was to keep away the Euloowayi, black fellows possessed of devils, who came from behind the sunset.

The other was a plain red-painted, tapering pine-pole which she said, when it fell to the ground, would tell of the death of some one related to an inmate of the house. Should it lean towards the house it foretold misfortune; or if she were any time away, when she was returning she would send her Mullee Mullee to sit on the top and bend it just to let us know. This pole would also keep away the spirits of the dead from the house during her absence. While she was away there would be no one to come and clear the place of evil by smoking the Budtha twigs all round it, as she always did if I were alone and, she thought, in need of protection.

Old Bootha has what she calls a wi-mouyan, clever-stick. It is about six feet long, great lumps of beefwood gum making knobs on it at intervals; between each knob it is painted. Armed with this stick, a piece of crystal, some green twigs, and sometimes a stick with a bunch of feathers on top, and a large flat stone, she goes out to make rain. The crystal and stone she puts under the water in the creek, the feathered stick she erects on the edge of the water, then goes in and splashes about with green twigs, singing all the time.

After a while she gets out and parades the bank with the wi-mouyan, singing a rain-song which charms some of the water out of the creek into the clouds, whence it falls where she directs it. Once my garden of roses looked very wilted. I asked Bootha to make rain, but just then she was very offended with Matah. One of her dogs had been poisoned, she would make no rain on his country. However, at last she said she would make some for me. I bound her down to a certain day. The day came; a heavy storm fell just over my garden, filling the ground tank, which was almost empty. About two inches fell. Within half a mile of each side of the garden the dust was barely laid.

Old Bootha's luck stuck to her that time, and I had to give her a new dress and some 'bacca.' But during the last drought she failed signally. Her excuse for failing was that a great wirreenun up the creek was so angry with the white people who were driving away all emu, kangaroo, and opossums, the black fellow's food, and yet made a fuss if their dogs killed a sheep for them sometimes, that he put his rain-stone in a fire, and while he did that no rain would fall. He said if all the sheep died the white fellows would go away again, and then, as long ago, the black fellows' country would have plenty of emu and kangaroo.

We saw a curious coincidence in connection with one of Bootha's witch-poles in my garden, the pole whose falling foretold death of some relative of some one in the house.

One afternoon there had been drizzling rain and a grey mist overshadowing things. Matah went out to look at the chances of a continuance of rain, the usual drought being on. He called to me to come and see a curious sky. Looking towards the west I saw a golden ball of a sun piercing the grey clouds which seemed like a spangled veil over its face; shooting from the sun was a perfect halo of golden light, from which three shafts spread into roadways up past the grey clouds into the vault of heaven. The effect was very striking indeed, against the grey clouds shaded from silver to almost black.

As we stood waiting for the sun to sink and the afterglow to paint these clouds, as it did, from shrimp pink and heliotrope to vivid crimson, we saw Bootha's pole fall. The air was quite still.

'The damp has loosened its setting,' said Matah, 'but we had better leave it alone and let the old girl fix it up again herself; it may be taboo to ordinary mortals like us.'

We left it.

That evening a messenger arrived from the sheep station to say my cook's mother had died just before sunset. The camp were firm believers in Bootha's witch-stick after that.

It was just as well we did not touch that stick; had we done so, Bootha says we should have broken out in sores all over our bodies.

They say that long ago the wirreenuns always used to have a sort of totem wizard-stick guarding the front of their camps.

BIRTH--BETROTHAL--AN ABORIGINAL GIRL FROM INFANCY TO WOMANHOOD

To begin at the beginning, Bahloo, the moon, is a sort of patron of women. He it is who creates the girl babies, assisted by Wahn, the crow, sometimes.

Should Wahn attempt the business on his own account the result is direful; women of his creating are always noisy and quarrelsome.

Bahloo's favourite spot for carrying on the girl manufacturing is somewhere on the Culgoa. On one of the creeks there is to be seen, when it is dry, a hole in the ground. As water runs along, the bed of this creek, gradually a stone rises from this hole. As the water rises it rises, always keeping its top out of the water.

This is the Goomarh, or spirit-stone, of Bahloo. No one would dare to touch this stone where the baby girls' spirits are launched into space.

In the same neighbourhood is a clear water-hole, the rendezvous of the snakes of Bahloo. Should a man go to drink there he sees no snakes, but no sooner has he drunk some of the water than he sees hundreds; so even water-drinkers see their snakes.

The name of the hole is Dahn.

Spirit-babies are usually despatched to Waddahgudjaelwon and sent by her to hang promiscuously on trees, until some woman passes under where they are, then they will seize a mother and be incarnated. This resembles the Arunta belief, but with the Euahlayi the spirits are new freshly created beings, not reincarnations of ancestral souls, as among the Arunta. To live, a child must have an earthly father; that it has not, is known by its being born with teeth.

Wurrawilberoo is said to snatch up a baby spirit sometimes and whirl along towards some woman he wishes to discredit, and through the medium of this woman he incarnates perhaps twins, or at least one baby. No doubt were it not for signs of teeth in a spirit-baby of immaculate conception, many a camp scandal would be conveniently nipped in the bud.

Babies are sometimes sent directly to their mothers without the Coolabah-tree or whirlwind medium.

The bronze mistletoe branches with their orange-red flowers are said to be the disappointed babies whose wailing in vain for mothers has wearied the spirits who transform them into these bunches, the red flowers being formed from their baby blood. The spirits of babies and children who die young are reincarnated, and should their first mother have pleased them they choose her again and are called millanboo--the same again.

They can instead, if they like, choose some other woman they know, which seems very accommodating in those presiding over the reincarnation department.

Sometimes two baby spirits will hang on one branch and incarnate themselves in the same woman, who as result is the mother of twins, and the object of much opprobrium in the camp. In fact, in the old days, one of the twins would have been killed.

One of my Black-but-Comelys said, on hearing that a woman had twins:

'If it had been me I would have put my fingers round the throat of one of them and killed it.' The woman who made this speech I had always looked upon as the gentlest and kindliest of creatures.

The father of the twins has treated his wife with the utmost contempt since their birth, and declines to acknowledge more than one of the babies.

They say the first-born of twins is always born grinning with his tongue out, as if to say, 'There's another to come yet; nice sort of mother I have.'

No wonder the women cover themselves under a blanket when they see a whirlwind coming, and avoid drooping Coolabah trees, believing that either may make them objects of scorn as the mother of twins.

When a baby is born, some old woman takes the Coolabah leaf out of its mouth. Such a leaf is said always to be found there if the baby was incarnated from a Coolabah tree; should this leaf not be removed it will carry the baby back to spirit-land. As soon as the leaf is taken away the baby is bathed in cold water. Hot gum leaves are pressed on the bridge of its nose to ensure its flatness; the more bridgeless the nose the greater the beauty.

When a baby clutches hold of anything as if to give it to some one, the bargie--grandmother--or some elderly woman takes what the baby offers, and makes a muffled clicking sort of noise with her tongue rolled over against the roof of her mouth, then croons the charm which is to make the child a free giver: so is generosity inculcated in extreme youth. I have often heard the grannies croon over the babies:

> Oonahgnai Birrablee,
> Oonahgnoo Birrahlee,
> Oonahgnoo Birrahlee,
> Oonabmillangoo Birrahlee,
> Gunnoognoo oonah Birrahlee.

Which translated is:

> 'Give to me, Baby,
> Give to her, Baby,
> Give to him, Baby,
> Give to one, Baby,
> Give to all, Baby.'

As babies are all under the patronage of the moon, the mothers are very careful every new moon to make a white cross-like mark on the babies' foreheads, and white dabs on cheeks and chins.

And very careful are the mothers not to look at the full moon, nor let their babies do so; an attack of thrush would be the result.

Bahloo, too, has a spiteful way of punishing a woman who has the temerity to stare at him, by sending her the dreaded twins.

If babies do not sleep well their mothers get the red powdered stuff like pine pollen, from the joints of the Bingahwingul, or needlebush tree, and rub it on the babies' skulls and foreheads.

If the babies cry too much their mothers say evil spirits are in them, and must be smoked out. They make a smoke fire of Budtha twigs and hold the baby in the thick of the smoke. I have seen the mother of a fretful child of three or four years even, apply the smoke anodyne.

Whenever the mother of a young child woke in the night, if well up in her mother duties, she was supposed to warm her hands, and rub her baby's joints so that the child might grow lissome and a good shape, and she always saw that her baby's mouth was shut when the child was asleep lest an evilly disposed person should slip in a disease or evil-working spirit. For the same reason they will not let a baby lie on its back unless they cover its head.

If a gilah flies over the camp crying out as it passes, it is a sure sign of 'debbil debbil'; the child, to escape evil consequences, must be turned on to its left side.

If a gooloo, or magpie, did the same, the child had to be laid flat on her moobil--stomach: for the passing of a cawing crow, a child had to be laid on the right side.

As these birds are not night birds, it is evident that they are evil spirits abroad in bird form, hence the precautions. As soon as a baby begins to crawl, the mother finds a centipede, half cooks it, takes it from the fire, and catching hold of her child's hands beats them with it, crooning as she does so:

> 'Gheerlayi ghilayer,
> Wahl munnoomerhdayer,
> Wahl mooroonbahgoo,
> Yelgayerdayer deermuldayer,
> Gheerlayi ghilayer.'

Which means:

> 'Kind be,
> Do not steal,
> Do not touch what to another belongs,
> Leave all such alone,
> Kind be.'

The accompaniment being a muffled click of a rolled-up tongue against the roof of a mouth.

No child must touch the big feathers of a goomblegubbon, or bustard's wings, nor any of its bones. At the age of about four, the mother takes one of these wings and beats the child all over the shoulders and under the arms with it. Again making the clicking noise, she croons:

> 'Goobean gillaygoo,
> Oogowahdee goobolaygoo,
> Wahl goonundoo,
> Ghurranbul daygoo.'

Which charm means:

> 'A swimmer be,
> Flood to swim against,
> No water,
> Strong to stop you.'

And so was a child made a good swimmer.

The wirreenuns would see that the septum of a child's nose was pierced at the right time, and their tribal marks cut on them. The nose was pierced at midwinter when ice was about, with which to numb the place to be pierced; ice was held to the septum, then prod through it went a bone needle.

An old gin who worked about the station had a pierced nose, and often wore a mouyerh, or bone, through it. A white laundress wore earrings. She said one day to the old gin:

'Why you have hole made in your nose and put that bone there? No good that. White women don't do that.'

The black woman looked the laundress up and down, and finally anchored her eyes on the earrings.

'Why you make hole in your ears? No good that. Black gin no do that, pull 'em down your ears like dogs. Plenty good bone in your nose make you sing good. Sposin' cuggil--bad--smell you put bone longa nose no smell 'im. Plenty good make hole longa nose, no good make hole longa ears, make 'em hang down all same dogs.' And off she went laughing, and pulling down the lobes of her ears, began to imitate the barking of a dog.

There is often a baby betrothal called Bahnmul.

For some reason or another it has been decided that a baby girl is to be given to a man, perhaps because he has been kind to her mother, perhaps she is owed to his kin by her own; any way the granny of the baby girl puts feathers, white swansdown, on the baby's head, and takes her over to the man when she is about a month old. Granny says to the baby:

'Look at him, and remember him, because you are promised to him.'

Then she takes some feathers off the baby's head and puts them on to his; that makes it a formal betrothal, binding to both sides.

I have heard great camp rows because girls made a struggle for independence, having found out they had only been promised, not formally betrothed, to some old chap whom they did not wish to marry. Perhaps the old fellow will already have a wife or so, a man can have as many as he pleases. I have heard of one with three; I have known some with two; but the generality of them seem content with one.

Should a young girl marry a man with an old wife, the old wife rules her to any extent, not even letting her have a say about her own children, and no duenna could be stricter. Should the young wife in the absence of her husband speak to a young man, she will probably get a scolding from the old wife and a 'real hiding' from the old man, to whom the old wife will report her conduct. Quite young men often marry quite old women; a reason sometimes given is that these young men were on earth before and loved these same women, but died before their initiation, so could not marry until now in their reincarnation.

Certainly, amongst the blacks, age is no disqualification for a woman; she never seems to be too old to marry, and certainly with age gains power.

At whatever age a girl may be betrothed to a man he never claims her while she is yet Mullerhgun, or child girl; not until she is Wirreebeeun, or woman girl.

A girl's initiation into womanhood is as follows. Her granny probably, or some old woman relation, takes her from the big camp into the scrub where they make a bough shade. As soon as this is made, the old woman sets fire to a thick heap of

Budtha leaves and makes the girl swallow the smoke. She then bids her lie down in a scooped-out hollow she has made in the earth, saying to her, 'You are to be made a young woman now. No more must you run about as you please. Here must you stay with me, doing as I say. Then in two moons' time you shall go and claim your husband, to do for ever what he bids you. You must not sleep as you lie there in the day time, nor must you go to sleep at night until those in the camp are at rest. I will put food ready for you. Honey you must not eat again for four moons. At first streak of day you must get up, and eat the food I have placed for you. Then when you hear a bird note you must shake yourself all over, and make a noise like this.'

And the old woman makes a ringing noise with her lips.

'That you must do every time you hear a fresh bird note; so too when you hear the people in the camp begin to talk, or even if you hear them laugh or sneeze. If you do not, then grey will your hair be while you are yet a young woman, dull will your eyes be, and limp your body.'

Girls have told me that they got very tired of being away with only the old woman for so long, and were glad enough when she told them they were to move to a new camp, nearer to the big one, which the women had prepared for them.

When they reached this the old woman rubbed off the mud with which she had plastered the girl's limbs when first they went away to camp, and which she had renewed from time to time. When this was all off she painted the girl in different designs with red ochre and white gypsum, principally in spots. She put on her head a gnooloogail, or forehead band, made of Kurrajong fibre, plaited and tied with some Kurrajong string, from over the cars to the back of the head; in this band, which she had painted white, she stuck sprays of white flowers. Sweetly scented Budtha and clustering Birah were the flowers most used for this ceremony. Should neither of these be in bloom, then sprays of Collarene or Coolibah blossom were used. When the flowers were placed in the band the old woman scattered a handful of white swansdown over the girl's head. Next she tied round her a girdle of opossum's sinews with strands of woven opossum's hair hanging about a foot square in front. Round her arms she bound goomils--opossum hair armlets--into which she placed more sprays of flowers, matching those in the girl's hair.

To show that the occasion was a sacred one a sprig of Dheal tree was placed through the hole in the septum of the nose. The toilet of a wirreebeeun was now complete.

The old woman gave her a bunch of smoking Budtha leaves to carry, and told her what to do. Note here the origin of bridal bouquets.

Having received her instructions, the girl, holding the smoking twigs, went towards the big camp.

When the women there saw her coming they began to sing a song in, to her, a strange language.

On a log, with his back towards her--for he must not yet look on her face--sat the man to whom she was betrothed. The girl went up to him. As the women chanted louder she threw the smoking Budtha twigs away, placed a hand on each of his shoulders and shook him. Then she turned and ran back to her new camp, the women singing and pelting her with dry twigs and small sticks as she went. For another moon she stayed with her granny in this camp, then the women made her another one nearer.

In a few weeks they made her one on the outskirts of the main camp. Here she stayed until they made her another in the camp, but a little apart. In front of the opening of this dardurr they made a fire. That night her betrothed camped on one side of this fire and she on the other. For a moon they camped so. Then the old granny told the girl she must camp on the same side of the fire as her betrothed, and as long as she lived be his faithful and obedient wife, having no thought of other men. Should he ill-treat her, her relations had the power to take her from him. Or should he for some reason, after a while, not care for her, he can send her back to her people; should she have a child he leaves it with her until old enough to camp away from her, when it is returned to him.

The wedding presents are not given to the bride and bridegroom, but by the latter to his mother-in-law, to whom, however, he is never allowed to speak. Failing a mother-in-law, the presents are given to the nearest of kin to the wife. You can hardly reckon it as purchase money, for sometimes a man gives no presents and yet gets a wife.

In books about blacks, you always read of the subjection of the women, but I have seen henpecked black husbands.

There are two codes of morals, one for men and one for women. Old Testament morality for men, New Testament for women. The black men keep the inner mysteries of the Boorah, or initiation ceremonies, from the knowledge of women, but so do Masons keep their secrets.

As to the black women carrying most of the baggage on march, naturally so; the men want their hands free for hunting en route, or to be in readiness for enemies in a strange country.

Black women think a great deal of the Moonaibaraban, or as they more often call them, Kumbuy, or sister-in-law. These are spirit-women who come a few days after the Boorah to bring presents to the women relations of the boys who have been initiated. The Kumbuy are never seen, but their voices are heard-voices like dogs barking; on hearing which the women in the camp have to answer, calling out:

'Are you my Kumbuy?'

An answer comes like a muffled bark, 'Bah! bah bah!'

Then the old men-crafty old men-go out to where the 'bahing' comes from, and bring in the gifts, which take the form of food, yams, honey, fruit principally.

These Kumbuy are among the few beneficent spirits they never hurt any one, simply supply the bereaved women with comfort in the shape of food, for the temporary loss of their male relatives. Should an uninitiate have a wife, which of course is improper, the Kumbuy decline to recognise her; and should she presume to answer their spirit back, they make in token of displeasure a thudding noise as if earth were being violently banged with a yam stick. She has encroached on the Kumbuy preserves, for prior to his initiation a man should only have a spirit wife, never an incarnate one.

If you ask a black woman why the Kumbuy thud the earth in answer to an initiate's wife, she will say:

'Dat one jealous.' jealousy even in the spirit world of women!

Unchaste women were punished terribly. After we went west even the death penalty for wantonness was enforced, though at the time we did not know it.

Should a girl be found guilty of a frailty, it being her first fault, her brothers and nearest male relations made a ring round her, after having bound her hands and feet, and toss her one from the other until she is in a dazed condition and almost frightened to death.

The punishment over, she is unbound and given to her betrothed, or a husband chosen for her.

Should a woman have been discovered to be an absolute wanton, men from any of the clans make a ring round her, she being bound, and tossed from one to the other, and when exhausted is unbound and left by her relations to the men to do as they please to her--the almost inevitable result is death. With this terror before them, it is possible the old blacks are right who say that their women were very different in their domestic relations in olden times.

THE TRAINING OF A BOY UP TO BOORAH PRELIMINARIES

AT the boy manufactory, Boomayahmayahmul, the wood lizard, was the principal worker, though Bahloo from time to time gave him assistance.

The little blacks throw their mythical origin at each other tauntingly. A little black girl, when offended with a boy friend, says:

'Ooh, a lizard made you.'

'Wah! wah! a crow made you,' he retorts.

Up to a certain age boys are trained as are girls--charms sung over them to make them generous, honest, good swimmers, and the rest; but after that they are taken into the Weedegah, or bachelors' camp, and developed on manly lines.

When he is about seven years old, his mother will paint her son up every day for about a week with red and white colourings. After that he would go to the Weedegah Gahreemai, bachelors' camp. He would then be allowed to go hunting with boys and men. He would see, now when he was out with the men, how fire was made in the olden time, almost a lost art now when wax matches are plentiful.

No boy who had not been to a Boorah would dare to try to make fire.

The implements for fire-making are a little log about as thick as a man's arm, of Nummaybirah wood--a rather soft white wood-- and a split flat piece about a foot long and three inches wide. The little log was split open at one end, a wedge put in it, and the opening filled up with dry grass broken up. This log was laid on the ground and firmly held there; the fire-maker squatted in front, and with the flat piece rubbed edgeways across the opening in the log. The sawdust fell quickly into the opening. After about a minute and a half's rubbing a smoke started out. After rubbing on a little longer the fire-maker took a handful of dry grass, emptied the smoking sawdust and dry grass into it, waved it about, and in three and a half minutes from starting the process I have seen a blaze. Sometimes it has taken longer, but just under five minutes is the longest time I have ever seen it take.

They use pine too, I believe, but whenever I timed them it was Nummaybirah they were using.

The boys pick up the woodcraft of the tribes when they begin going out with the men. As the boys began to grow up, when a good season came round, and game and grass were plentiful, the old men were seen to draw apart often and talk earnestly.

At length there came a night when was heard a whizzing, whirling boom far in the scrub. As the first echo of it reached the camp, the women, such as were still young enough to bear children, stopped their ears, for should any such hear the Gurraymi, the women's name for the Gayandi, or Boorah spirit's voice, that spirit will first make them mad, then kill them.

The old women began to sing a Boorah song. To deaden the sound of the dreaded voice, opossum rugs were thrown over the children, none of whom must hear, unless they are boys old enough to be initiated; the sound reveals the fact to such that the hour of their initiation is at hand.

The men all gathered together with the boys, except two old wirreenuns, who earlier in the evening have seemingly quarrelled and gone away into the scrub.

The men and boys in camp march up and down to some distance from the camp. The old women keep on singing, and one man with a spear painted red with a waywah fastened on top, walks up and down in the middle of the crowd of men, holding the spear, with its emblematic belt of manhood, aloft; as he does so, calling out the names of the bends of the creek, beginning with the one nearest to which they are camped. When he gets to the end of the names along that creek and comes to the name of a big river, all the men join him in giving a loud crow like

'Wah! wah! wah!'

Then he begins with the names along the next creek across the big river, and so on; at the mention of each main stream the crowd again join in the cry of

'Wah! wah! wah!'

All the while, closer and still closer, comes the sound of the Gayandi, as the men call the Gurraymi, or bull roarer.

At length the two old wirreenuns come back to the camp and the noise ceases, to recur sometimes during the night, when I expect, did any one search for them, the old wirreenuns would be found missing from the camp.

After the first whirling of the bull roarers and calling of the creek names, the Gooyeanawannah, or messengers, prepare for a journey, and when ready, the wirreenuns start them off in various directions to summon neighbouring tribes from hundreds of miles round to attend the Boorah. The messengers each carry a spear with a waywah (or belt of manhood) on the top, seeing which no tribe, even at enmity with the messenger, will molest him. When a messenger arrived at a strange camp, he was not asked his business but left to choose his own time for telling. He would squat down a little way from the strangers' camp, food would probably be brought to him which he would cat.

He would find out who was the chief wirreenun of the tribe, then take him apart, give to him his Boorah message-stick as guarantee of his good faith, and tell him where and when the Boorah was to be held. After having given all necessary information, the Gooyeanawannah would return to his tribe; the wirreenun to whom he had given the Doolooboorah, or message-stick, would send it on by the messenger of his tribe, and so with others, until all were summoned, each tribe letting it be known that a Boorah summons had been received by sounding the Gayandi, which would carry its own tale to those in the camp.

Should young boys be chosen as messengers, they were held in high honour; Woormerh they were called.

While the messengers were away, the old men of the tribe in whose Noorumbah, or hereditary hunting lands, the Boorah was to be held, prepared the sacred grounds.

They cleared a big circle, round which they put a bank of earth, and from the circle was cleared a path leading to a thick scrub;

along this path were low earthen embankments, and the trees on both sides had the bark stripped off, and carved on them the various totems and multiplex totems of the tribes. Such carvings were also put on the trees round the Bunbul, or little Boorah ring, where the branches were also in some instances lopped, and the trunks carved and painted to represent figures of men, amongst whom were supposed to be the sons of Byamee's wives. Two of these sons had been made young men at the first Boorah Byamee instituted in this district, the ground of which is pointed out to this day.

In the middle of the Bunbul a large heap of wood was placed ready for the Yungawee, or sacred fire.

When the preliminary preparations were over, the camps were moved to just outside the Boorah, or big Boorah ring. By that time the other tribes began to arrive. First came from each tribe the boys to be initiated and the Munthdeeguns, or men in charge of them. The men were painted, and had leafy twigs tied round their wrists and ankles, as had the boys also, and all carried in their hands small branches of green. Those especially in charge of boys held, too, a painted spear with a waywah on top of it.

As they approached the place of gathering the head man, with the painted spear, began calling out all the names of the places along the creeks from whence he came; at the name of each big watercourse they all cried together

'Wah! Wah! wah!'

They were met at some distance from the camp by the men who had summoned them, and who had made a round brush yard where they were to meet them. Here the older women were singing Boorah songs. Some held their breasts as a sign they had sons among the initiates; others put their hands on their shoulders, which showed they had brothers going to be made young men. All the women had leafy twigs tied round their wrists and ankles as the men had. The newcomers and the men who met them walked round the yard at a measured beat, lifting one leg and throwing up one arm each time the cry of 'Wah! wah! wah!' was given, for here too the enumeration of geographical names went on.

When the Boorah song was over, the men marched out of the yard; closely behind them the two oldest men with the tufted spears; the Boorah boys closely after them. The women followed, carrying bunches of leafy twigs with which they pelted the boys until they reached the camp.

Matah and I had been watching the whole performance, and followed in the wake of the women.

The whole scene impressed us as picturesque-the painted figures of the men and boys, with the peculiarly native stealthy tread, threading their way through the grey Coolabah trees; the decorated women throwing their leafy missiles with accurate aim into the ranks of the boys, who did not dare to look at their assailants. A Boorah boy must give no evidence of curiosity; the nil admirari attitude then begun clings to a black man through life. The women of the tribe express voluble surprise, but a black man never except by the dilation of his eyes.

Every night after this a corroboree was held. The fully initiated of each tribe, as they arrived, help in the preparation of the inner sacred ground, while the younger men collected game and other food.

The old men cut out of the ground along the narrow path leading to the Bunbul, and round it, huge earthen animals, their various totems, such as crocodiles, kangaroos, emus and others, all of a colossal size. These they plastered over with mud and painted in different colours and designs. On the right of the Bunbul they made an earthen figure of Byamee--this figure was reclining holding in each hand a Boondee. On the other side was the huge figure of a woman-this represented Birrahgnooloo, the favourite wife of Byamee; she held two spears. There was a third figure not so large as the other two but like them, apart from the figures near the path and the Bunbul; this was Baillahburrah, according to some, Dillalee according to others, the supernatural son of Byamee--or as some say, brother--not born of woman, having lived before the human race existed, before Byamee travelled as Creator and culture hero through Australia.

Of the Gayandi, the Boorah spirit, sometimes called Wallahgoo-roonboooan, there was no figure, because he was always present at Boorahs, though invisible. His voice only gave evidence of his presence.

The wirreenuns said it was he who had placed in the forks of trees round the big ring heaps of dry wood, which they said, when the ceremonies began, he would light, making a dazzling illumination of the scene.

In the middle of the Boorah ring was placed a mudgee, a painted stick on- spear, with a bunch of hawk's feathers on the top. Every night was heard at intervals the Gayandi, and immediately the younger women and children stopped their ears, while the old women shrieked their brumboorah.

As each fresh batch of blacks arrived the volume of sound was increased, for the old men with their Gayandi would go into the scrub and whirl them. These bull roarers sound curiously uncanny-I did not wonder the uninitiated accepted the spirit theory as to their origin.

The bush of Australia is a good background for superstition; there is such a non-natural air about its Nature, as if it has been sketched in roughly by a Beardsley-like artist.

The function of the Gayandi is to inspire awe, and it fulfils it. Byamee himself made the first. It was some time before he got quite the effect he wanted. At first he desired to give the Boorah spirit a form as well as a voice, to inspire awe; he also wished it to knock out the front tooth of an initiate.

He made a stone figure in the image of man, having a voice. This spirit, known variously as Gayandi, or Darramulun, went to the Boorah, but when he was to knock out the front tooth, he began to eat the boys' faces. He was too strong; he would not do to preside over, Boorahs. Byamee transformed him into a large piggiebillah-like animal, though instead of being covered with spines, thick hair grew over him; he has since been known as Nahgul. He went away into the bush, where he has been a dreaded devil ever since; for if he touches a man's shadow even, that man will itch all over and nothing can cure him of it. He haunts Boorah grounds.

Next Byamee made a stone bull roarer sort of thing, but this was too heavy to make the noise he wanted. One day he was chopping a big Coolabah tree close to Weetalibah water-hole, which tree, much to the horror of our blacks, was burnt down a few years ago by travellers.

As Byamee chopped, out flew a big chip. He heard the whizzing sound it made, gave another chop, out flew another; again the whizzing sound.

'That is what I want,' he said I'll make a Gayandi of wood.'

He cut a piece of mubboo, or beefwood, and shaped it; he tied a piece of string to a hole in one end; he hung it up in the big Coolabah tree. Then he went and cut one out of Noongah or Kurrajong, tied a string on to that and put it beside the other on the tree, and left them swinging there.

One day he came back and was camping near; his wives, came along to the big tree. There the Gayandi swung, making a whirring noise.

'What's that?' said the women. 'We'll have a look what it is.' Seeing Byamee they said, 'We heard voices in that big tree over there.'

Whereabouts?' he said.

In that Coolabah tree. Such strange voices, such as we never heard.'

'You two go" he said, 'to our camp and make a fire. I'll go and see what it is.'

When the women were out of sight he went to the tree and took the pieces of wood down. He was satisfied now they would answer his purpose. He carefully hid them until he made a Boorah. And since then such pieces of wood have been the medium for the Gayandi's voice, and are kept carefully hidden away from all but the eyes of wirreenuns.

At length all the expected tribes had arrived, preparations were finished, and a signal was given for a move to be made that the real ceremonies might begin.

The fully initiated men went away after their midday meal, and about sundown came in single file along the banked-in path each carrying a firestick in one hand, a green switch in the other. When they reached the mudgee in the middle of the big ring and corroboreed for a little round it, the old women answered with a Boorah song, and all moved to the edge of the ring. At this stage men often tried to steal each other's boys, and great wrestling matches came off. One man would try to pull up the mudgee, out would rush one of another clan to wrestle with him. First the boys would wrestle, then the elder men, each determined his clan should prove victorious at this great Boorah wrestling.

The skill of the eeramooun, or uninitiated boys, would be tried in sham fights too. They were given bark shields, and their attackers had bark boomerangs; great was the, applause when the boys ably defended themselves. Previously they have been tried with boomerang and boodthul throwing, and other arts of sport and warfare, boys of each tribe trying to excel those of the others. If a boy comes well out of these trials the men say he is worthy to be a yelgidyi, or fully initiated young man.

When the wrestling and sham fights are over, corroborees begin. All night they are kept up, and sometimes there are day performances too.

THE BOORAH AND OTHER MEETINGS

AT last would come the night when everything was ready. Sports and corroborees would be held as usual, until, at a given signal, the younger women were ordered into bough sheds which were round the ring.

The old women stayed on singing.

The boys, who are painted red, are beckoned into the middle of the ring, where their respective Munthdeeguns daub them with white. That done, each man seizing his charge, hoists him on to his shoulder, and dances round the ring with him. Then the old women are told to bid the boys good-bye.

Forward they come, singing each her own brumboorah, for every oldest woman relation of each of the boys makes a song for him. They corroboree a few steps behind the men, chanting a farewell, then corroboree back a few steps, then hasten to join the younger women in the bough sheds, which are now pulled down on top of them by the men, that they may see nothing further. Then the Munthdeeguns disappear down the track into the scrub.

When they are out of sight the women are released, that they may get ready to travel to where the Durrawunga, or Little Boorah, will be held in about four days' time, at about ten miles distance.

As the Munthdeeguns passed their totem-marked trees, or images, which would be those of the boys in their charge--for each guardian was a relation of the same totem as his charge-they would perform some magical feat, such as producing gubberahs, charcoal, gypsum, and so on, uttering as they did so a little chant about that totem.

The boy's eyes are closed all this time and his head bent down.

Boys at a Boorah always remind me of Wilhelm Meisler's Travels, where, at the school to which Wilhelm takes Felix, he learns, on inquiry as to the three attitudes assumed by the pupils, that these gestures inculcate veneration, which also seems to be the keynote of the eeramooun's instruction. The Boorah over, he too, 'Stands erect and bold, yet not selfishly isolated; only in an union with his equals (his fellow initiates) does he present a front towards the world.'

And only when the fear, the abasement, is gone does the true reverence come, which makes the most primitive creed a living religion.

As the Munthdeeguns pass the sacred fire they throw in a weapon each. This done they place their charges in slightly scooped-out places, already prepared in the inner ring.

Then they bid them, on pain of death, not to look up whatever happens.

Soon a great whirring is heard, telling that Gayandi, the Boorah spirit, is near.

Yudtha Dulleebah, one of the oldest black men in the district, said at this stage once two boys did look up.

The wirreenuns saw them, though the boys did not know it and went on looking. These boys saw the men advance each to the fire where they had thrown their weapons; chanting in a strange tongue, they corroboreed round the fire for some time.

Then the wirreenuns snatched up the coals left from the weapons and rubbed them into their limbs, trampling as they did so on the edge of the fire, which did not seem to burn them, rubbing and chanting until the sacred coals were supposed to be absorbed by them, from which they would derive new powers.

This over, the boys were all ordered to get up, and march round, hands on thighs and heads abased, while they learnt a Boorah song, giving new words for common things, which acted as pass-words hereafter for the initiated. Into a slow chant these words were strung, as the men and boys passed round the ring, two of the oldest men standing beating time with painted spears with tufted tops.

The two boys who had transgressed before looked up again, curious as to their surroundings. Suddenly the men with the spears roared at the boys to lower their heads.

The boys laughed. Their fates were sealed. Out flashed the sacred gubberahs of these two old men.

'Dead is he,' they cried, 'who laughs in the Bunbul where yungawee burns more fiercely than Yirangal, the sun, where near lies the image of Byamee: Byamee, father of all, whose laws the tribes are now obeying.' Then the men chanted to the gubberahs and held them between the fires and the boys, the light of the flames seemed to play on them and stretch its beams to the boys, who began to tremble. As louder grew the chant an answer came from the scrub, the voice of Gayandi;

shaking with fear the boys fell to the ground, to all appearance lifeless. Then the old men went forward, each with a stone knife in hand. Stooping over the two boys they opened veins in each, out flowed the blood, and the other men all raised a death cry. The boys were lifeless. The old wirreenuns, dipping their stone knives in the blood, touched with them the lips of all present. Then the bodies were put on the edge of the sacred fire and the other initiates taken a little further into the scrub. There they were tried in many ways.

With the Boorah spirits whistling and whizzing all round them, spears were pointed at them. Their skins were scratched with stone knives and mussel shells. Hideously painted, fiendish-looking creatures suddenly rushed upon them. Should they show fear and quail at the Little Boorah they would be returned to their mothers as cowards unfit for initiation, and sooner or later sympathetic magic would do its work, a poison-stick or bone would end them. Or if one of the initiates was considered stupid and generally incapable, having been brought to the Boorah for that purpose, he was now, after having been made to suffer all sorts of indignities, such as eating filth and so on, bound to the earth, strapped down, killed, and his body burnt.

When the trials were over and the old wirreenuns said to the boys who had not quailed, 'You are brave; you shall be boorahbayyi first and afterwards yelgidyi, and carry the marks that all may know.'

Then they made on the shoulder of each boy a round hole with a pointed stone; this hole they licked to feel no splinter of stone remained, then filled it with powdered charcoal.

After this, leaving the boys there, the men went back to the Bunbul ring. The bodies of the Boorah victims were cooked.

Each man who had been to five Boorahs ate a piece of this flesh, no others were allowed even to see this done. Then the bones and what was left of the bodies were put into the middle of the fire, and all traces of the victims so destroyed.

The men then sang a song, saying that so must always be served those who scoffed at sacred things; that the strength they had wasted should go into other men who would use it better; while the spirits of the victims should wander about until reincarnated if the Boorah spirit gave them another chance. Perhaps he would only let them be reincarnated in animals.

After another dance and chant round the yungawee, the men went and brought the boys back again. They came with their hands on their thighs, and their heads abased; each was taken to his allotted place near the outer edge of the ring. There each Munthdeegun told his boy he could sleep that night; he would go to sleep the boy he had been, to wake in the morning a new man; his courage had now been tried, and in the morning a new name and a sacred stone would be given to him. The Gayandi would settle their names that night and tell the wirreenuns.

The next morning the boys were awakened by the Munthdeegun chanting and dancing before them. They stopped in front of the first boy, called him to rise by a new name; as he did so all the men clapped their thighs and shouted

'Wah! wah! wah!'

Then an old wirreenun gave him a small white gubberah, which he was bidden to keep concealed for ever from the uninitiated and the women, and he must be ready to produce it whenever called upon to do so. The result of failure would be fatal to him. With the loss of the stone his life spirit would be weakened, and the strength of the Boorah spirit, with which he was now endowed, be used against him instead of for him, as would be the case as long as he kept the stone.

These stones seem somewhat in the way of 'Baetyli' of pagan antiquity, which were of round form; they were supposed to be animated, by means of magical incantations, with a portion of the Deity; they were consulted on occasions of great and pressing emergency as a kind of divine oracle, and were suspended either round the neck or some other part of the body.

As each boy received his stone another loud chorus of 'Wah! wah! wah!' went up from that crowd, making the scrub ring with the sound.

Some of those, of whose tribe it was the custom-it is not invariably so-now had a front tooth knocked off; this done a wirreenun chanted to the boy, who had been blindfolded and almost deafened by the whirring of Gayandi.

One chant was as follows:--

> Now you can meet the Boorah spirit,
> Now will he harm you not.
> He will know his spirit is in you.
> For this is the sign,
> A front tooth gone.
> That is his sign,
> He will know you by it.'

Some of the wirreenuns buried these teeth by the Boorah fire, others carefully wrapped them up to keep as charms, or to send to other tribes, each according to the individual custom of his tribe.

This all over, once more there was a marching and chanting round the fire, then the boys were taken away and given food for the first time since they left their mothers.

No wonder that the 'supernatural' was mixed up with their impressions of the Boorah: fasting nourishes hallucinations. While the boys were eating, they could hear in the distance other chants, and knew that ceremonies were going on to which they were not yet to be admitted, there being degrees of initiation.

On the fourth day the men took them about ten miles, and camped with them where they could hear faintly in the distance the noise of the main camp; so they knew they were near the place chosen for the Durramunga, or Little Boorah.

Just before dawn next morning each Munthdeegun took his Boorahbayyi, or partially initiated one, to the Durramunga. There was a Boorah ring, but instead of earth grass was heaped all round it. No young women were visible, only the old women, who sang and corroboreed towards the boys. Slowly they came forward, peered at their shoulders, and seeing there the marks, embraced them, shrieking out cries of joy that their boys had borne the tests. They danced round them, then at a sign from the old men embraced them again; and while, the women sang their brumboorah and danced, the boys were taken away by their guardians.

For two moons they remained away, learning much as to sacred things. They were told that the oldest wirreenuns could see in their sacred crystals pictures of the past, pictures of what was happening at a distance in the present, and pictures of the future; some of which last filled their minds with dread, for they said as time went on the colours of the blacks, as seen in these magical stones, seemed to grow paler and paler, until at last only the white faces of the Wundah, or spirits of the dead, and white devils were seen, as if it should mean that some day no more blacks should be on this earth.

The reason of this must surely be that the tribes fell away from the Boorah rites, and in his wrath Byamee stirred from his crystal seat in Bullimah. He had said that as long as the blacks kept his sacred laws, so long should he stay in his crystal seat, and the blacks live on earth; but if they failed to keep up the Boorah rites as he had taught them, then he would move and their end would come, and only Wundah, or white devils, be in their country.

It is said that this prophetic vision was the reason that so many of the first-born half-caste babies were killed, the old wirree-nuns seeing in them the beginning of the end.

At the end of two moons they make back towards the place where the Boorah had begun, and where preparations were now being made to receive them.

They camped in the scrub near the old camp of the tribe who had started the Boorah.

That night in the camp the Gayandi was heard again, another ceremony was at hand.

The next day the women at the big camp made a big fire, a little distance away. When this fire was nearly burnt out they covered it thickly with Budtha, Dheal, and Coolabah leaves to make a great smoke. On the top of these leaves, which were piled about two feet high, logs were placed; this fire was round a Dheal tree.

When the thick smoke was seen curling up in a column, the Boorahbayyi were brought out of the scrub by the Munthdeegun, while in the distance sounded the whizzing voice of the Boorah spirit. As it ceased, when the women's chanting rose above it, the painted boys came into the open. On they came, heads down and hands on thighs, looking neither to the right nor to the left, but walking straight ahead until they stood on the logs on the fire. They leaned over and placed a hand each on the tree in the centre, there they stood while the smoke curled all round them. The women past child-bearing were singing all the time, while the men danced outside the leaf-smoke, clicking boomerangs as they did so.

For some time this went on, then the men took the boys back into the scrub.

In about four moons' time another leaf-smoke was made ready, and the Boorahbayyi were again brought out and smoked. This time while chanting a song the old women brought a big net and put it right over the boys. Then they stepped back and danced round to the clicking of boomerangs by the men. The boys were again taken away.

But after this they were allowed to camp nearer the general camp, though they held no intercourse with the people of it. I have often met these Boorah boys in the bush, and on sighting me they have fled as if I were a devil in petticoats.

In about another moon's time, the boys were painted principally white, a waywah put on them, a yunbean--a piece of beefwood gum with two kangaroo teeth stuck in it, and a hole through it-- was tied to their front lock of hair. A number of these yunbean were tied to forehead bands, which they wore too. Armlets of opossum's hair string were put on their arms, and feathers stuck in them. Feathers were also stuck upright in the forehead bands.

Some of the old men added to their own decorations by putting on wongins, from which were hanging those most precious possessions to inland blacks--seaside shells. Some had fresh beads of gum fastened on to their hair, hanging round their heads in dozens.

The women, too, had coiffured themselves with fresh gum beads; the mothers of the Boorahbayyi were painted, too, in corroboree style. They had made a smoke fire, but the logs instead of being put on it, were placed at a little distance; on these the painted boys sat, the smoke enveloping them.

After they had been seated there some time, their mothers came up behind them, and put their hands on their sons' shoulders. Then they rubbed all the paint off the boys' bodies; the boys never once looking at them. When the paint was all off, the women sang and danced, until the men in charge took the boys away again.

After this, supervision was relaxed except at night. During the day-time the boys might wander at will, so as they kept clear of the general camp. They might not receive food from nor speak to a woman for twelve months, as if they were monks of Byamee in training.

At his second Boorah a young man was allowed to see the sacred fire ceremony, throwing in of weapons, walking on burning coals, and the rest. He saw the huge earthen figures of Byamee, Birrahgnooloo, and Baillahburrah, or Dillalee, and was told all about them; that Byamee having initiated the Boorah, only such as have been through its rites can go to his sky-camp.

Three sins are unforgiveable, and commit a spirit of a guilty one to continual movement in the lower world of the Eleanbah Wundah, where, but for big fires kept up, would be darkness.

There the guilty one had to keep his right hand at his side, never moving it, but he himself perpetually moving. Those who know the blacks and their love of a 'dolce far niente,' will understand what a veritable hell this perpetual movement would make.

The three deadly sins were unprovoked murder, lying to the elders of the tribe, or stealing a woman within the forbidden degrees--that is, of the same hereditary totem, i.e. of the same blood, or of the prohibited family name clan.

But by a curious train of reasoning two wrongs make a right. Should by any chance a man succeed in getting a wife he had no right to, having lived with her, he could keep her, if he came unhurt from the trial he had to stand; he only having a shield to defend himself with, the men of the stolen woman's kin threw weapons at him. Only the men of her kin are assailants, not as in a murder trial, when the men of all kins can throw at the guilty man. Should he defend himself successfully, he can keep the woman on the understanding that a woman of his family is given to a man of hers, to square things. A man who stands his trial is called a Booreenbayyi.

Kindliness towards the old and sick is strictly inculcated as a command of Byamee, to whom all breaches of his laws are reported by the all-seeing spirit at a man's death, and he is judged accordingly. Sir Thomas Mitchell, writing in 1837 his experiences of the blacks during his explorations, notices as very striking their care and affection for the aged of their race.

At his second Boorah a man is allowed to see the carvings on the trees and to hear the legends of them. Also to hear the Boorah song of Byamee, which Byamee himself sang; and to hear the prayer of the oldest wirreenun to Byamee, asking him to let the blacks live long, for they have been faithful to his charge as shown by the observance of the Boorah ceremony.

The old wirreenun says words to this effect several times imploringly, his head turned to the east; facing this direction the dead are mostly buried.

Though we say that actually these people have but two attempts at prayers, one at the grave and one at the inner Boorah ring, I think perhaps we are wrong. These two seem the only ones directly addressed to Byamee. But perhaps it is his indirect aid which is otherwise invoked. Daily set prayers seem to them a foolishness and an insult, rather than otherwise, to Byamee. He knows; why weary him by repetition, disturbing the rest he enjoys after his earth labours? But a prayer need not necessarily be addressed to the highest god. I think if we really understood and appreciated the mental attitude of the blacks, we should find more in their so-called incantations of the nature of invocations. When a man invokes aid on the eve of a battle, or in his hour of danger and need; when a woman croons over her baby an incantation to keep him honest and true, and that he shall be spared in danger, surely these croonings are of the nature of prayers born of the same elementary frame of mind as our more elaborate litany. I fancy inherent devotional impulses are common to all races irrespective of country or colour.

When the prayer was over the old men chanted Byamee's song, which only the fully initiated may sing, and which an old black fellow chanted for us as the greatest thing he could do.

There seemed very little in this song, for no, one can translate it, the meaning having been lost in the 'dark backward,' if it was ever known to the Euahlayi.

> 'Byamee guadoun.
> Byamee guadoun.
> Byamee guadoun.
> Mungerh wirree.
> Mungerh wirree.
> Mungerh wirree.

> Birree gunyah, birrie gunyah.
> Dilbay gooran mulah bungarn.

> Oodoo doo gilah.
> Googoo wurra wurra.

> Bulloo than nulgah delah boombee nulgah.
> Delah boombee. Nulgah delah boombee boombee.
> Buddereebah . . . . . . Eumoolan.
> Dooar wullah doo. Boombee nulgah delah.'

The old fellow said wherever Byamee had travelled this song was known, but no one now knew the meaning of the whole, not even the oldest wirreenuns.

Another stone was given to a Boorahbayyi when he first heard this song.

The wirreenuns, they say, swallow their stones to keep them safe.

At each Boorah a taboo is taken off food. After a third Boorah a man could eat fish, after a fourth honey, after a fifth what he liked. He was then, too, shown and taught the meanings of the tribal message-sticks, and the big Boorah one of Byamee. As few men now have ever been to five Boorahs, few know anything about these last. At each Boorah a stone was given to a man, and when he had the five he could marry.

After each Boorah all the figures and embankments are destroyed.

After the fifth Boorah the mystery of the Gayandi was revealed and the bull roarers shown--oval pieces of wood pointed at both ends, fastened to a string and swung round; but though this was shown, the wirreenuns told them that the spirit's voice was really in this wood animating it. After a man has been to one Boorah he can have war weapons and is a warrior, but not until he has been to five can he join or be one of the dorrunmai-sort of chiefs-who hold councils of war, but have few privileges beyond being accepted authorities as to war and hunting. With the wirreenuns rests the real power, by reason of their skill in magic.

Besides Boorahs are minor corroboree meetings where marriages are arranged; meetings where the illegality of marriages is gone into, and, if necessary, exchanges effected or arranged; meetings where the wirreenuns of the Boogahroo produce the bags of hair, etc., and vendettas are sworn; meetings of Boodther, or giving, where each person receives and gives presents. A person who went to a Boodther without a goolay full of presents would be thought a very poor thing indeed.

Of course every meeting has a corroboree as part of it.

Every totem even has its own special corroboree and time for having it, as the Beewees, or iguanas, when the pine pollen is failing and the red dust-storms come. And if you abused these dust-storms to a Beewee black, you would insult him: it is not dust, it is the pollen off the pines, and so a multiplex totem to him!

The winds belong to various totems, and the rains are claimed by the totem whose wind it was that blew it up.

If a storm comes up without wind it belongs to Bohrah, the kangaroo.

The big mountainous clouds when they come from the south-west are said to be Mullyan, the eagle-hawk, who makes the south-west wind claimed by Maira, paddy melon totem, one of whose multiplex totems Mullyan is.

The crow keeps the cold west wind in a hollow log, as she was too fond of blowing up hurricanes; she escapes sometimes, but the crow hunts her back. But they say the log is rotting and she will get away yet, when there will be great wreckage and quite a change in climates.[28]

Away to the north-west a tribe of blacks have almost a monopoly in wind-making, holding great corroborees to sing these hurricanes up. One of this tribe came to the station once and wanted to marry a girl there. She would not consent, and told him to go home. He went, threatening to send a storm to wreck the station. The storm came; the house escaped, but stable, store, and cellar were unroofed. I told my Black-but-Comelys to kindly avoid such vehemently revengeful lovers for the future.

CHIEFLY AS TO FUNERALS AND MOURNING

I WAS awakened one morning on the station by distant wailing.

A wailing that came in waves of sound, beginning slowly and lowly, to gain gradually in volume until it reached the full height or limit of the human voice, when gradually, as it had risen, it fell again. No shrieking, just a wailing inexpressibly saddening to hear.

I lay for some minutes not realising what the sound was, yet penetrated by its sorrow. Then came consciousness. It was from the blacks' camp, and must mean death. Beemunny, the oldest woman of the camp, who for weeks had been ill, must now be dead.

Poor old Beemunny, who was blind and used to get her great-granddaughter, little Buggaloo, to lead her up to the tree outside my window, under whose shade she had spent so many hours, telling me legends of the golden age when man, birds, beasts, trees, and elements spoke a common language. But the day before I had been to the camp to hear how she was. The old women were sitting round her; one of the younger ones told me her end had nearly come.

The Boolees, or whirlwinds, with the Mullee Mullees of her enemies in, had been playing round and through the camp for days, they said, watching to seize her fleeting spirit-a sure sign the end was near. That night surely would come Yowee, the skeleton spirit, with the big head and fiery eyes, whose coming meant death.

Last night more than one of the blacks had dreamt of an emu, which meant misfortune to one of that totem, which was Beemunny's.

As Yellen spoke in a hushed sad voice, suddenly, though no breath of wind was stirring, sprang up on the edge of the camp a boolee, rearing its head as if it were a living thing. Round it whirled, snatching the dead leaves of the Coolabahs, swirling them with the dust it gathered into a spiral column, which sped, as if indeed a spirit animated it, straight to the camp of the dying woman. Round and round it eddied, a dust-devil dancing a dance of death.

The watchers drew nearer to Beemunny, who was past heeding even the spirits of evil.

The women in other camps clutched their children to them, but spoke no word. All was silent but the swirling leaves as the column gathered them. Finding the deathbed guarded, the boolee turned sharply from the camp and sped away down the road, dissolving on the poligonum flat in the distance.

Yellen gave a sigh of relief.

But now her fears were verified; Beemunny was dead.

Poor old Beemunny! How the vanities of youth cling to one; how we are 'all sisters under the skin.'

She was ever so old, she was blind, her face was scarred with wrinkles, yet one of her beauties remained, and she absolutely joyed in its possession: it was her hair. Her hair was thick and fuzzy, when combed would stand nearly straight out, which is quite unusual with the native women's hair in that part.

Beemunny one day asked one of the younger women if I had ever heard what a lot of lovers she had had in her youth, what fights there had been over her, and all because of her beautiful hair.

Poor old Beemunny! Something in my own woman nature went out to her in sympathy. She was old, she was ugly, her husband was dead, as were all men to her.

Poor old Beemunny! Having once learnt her vanity, I never passed her without saying 'Gubbah Tekkul!' 'Beautiful hair!' at which she would beam and toss her head.

At sunrise came again the wailing; the singing of the Goohnai, or dirge, wherein are enumerated all the multiplex totems of the deceased, crooned in a wailing way, and each fresh person who comes to the camp sings this dirge again. In olden times all would have been painted in full war paint, weapons in hand, to see the corpse.

I was given permission to go to the funeral, old Bootha was to take me.

I heard that Beemunny had died early in the night. Her daughter and nearest of kin had sat all night beside her body, with each a hand on it to guard her from the spirits. She was now in her bark coffin, round which were her own blankets to be buried with her. The coffin was made of bark cut off right round a tree, split on one side from end to end; the body was placed in this, then the bark lapped over it, the ends were blocked up with other pieces, the whole secured by ropes. All day until the burial some one of kin stayed beside the coffin, little fires of Budtha kept smoking all the while. In the afternoon old Bootha came for me, and we set out.

First in the procession marched two old men of the tribe, behind them some young men, then those in charge of the coffin and the two nearest women relations, immediately behind them the old women, then the young women. No women with babies were allowed to go, nor any children. I came last with old Bootha.

The procession moved along an old winding track on the top of a moorilla, or pebbly ridge, pine-trees overarching in places carving the sky into a dome-a natural temple through which we walked to the burial-ground.

Every now and then we heard a bird note, which made the women glance at each other and say, first, 'Guadgee,' then 'Bootha,' as it came again, and a third time 'Hippitha.' To my uneducated ear the note seemed the same each time. I asked Bootha what it was. She told me it was the note of a little bird, something like a wren, called Durrooee, in whose shape the spirits of dead women revisited the earth. It seems that Numbardee, the first woman, was, like Milton's Eve, a caterer; she acquired art in beating the roots of plants into flat cakes much esteemed; she was never to be met without some, carrying them always in a bag across her shoulders.

And Byamee was so pleased with her for always having food for the hungry that, when at length she died, he allowed her to revisit her old gahreemai, or camp, her spirit returning in the form of the little honey-eater bird, Durrooee; and all women after her had a like privilege if they had done their duty in life. These birds are sacred; no one must harm them, nor even imitate their cry. It would be hard to hurt them, for the spirit in them is so strong. If any one even takes up a stick or stone to throw at them, hardly is it raised from the ground when the would-be assailant is forcibly knocked over, though he sees nothing but the little bird he was about to attack. Then he knows the bird must be a spirit bird, and perhaps seeing him look at her, the bird calls a woman's name, then he knows whose spirit it is.

A black boy on the station was badly hurt by a fall from a tree. It had seemed strange that such a good climber should fall. The blacks said it was because there was a Durrooee's nest in that tree, the spirit had knocked him down, and for a time so paralysed the man with him that he could not move to his assistance. Needless to say, they have avoided that tree since.

In the distance we heard the sound of the grave being dug. None of the same totem as the dead person must dig the grave. The coffin was put down beside the grave, the daughter and other nearest women relations stayed with it, the other women went away into the bush in one direction, some of the men in another.

Old Hippi heaped up some Budtha twigs he had gathered, I noticed as we came along; these he set fire to, and made a dense smoke which hung low over the open grave and spread over the old graves.

Hippi smoked himself in this smoke. The women came back with arms full of small branches of the sacred Dheal tree, these they laid beside the grave, then sat down and broke them into small twigs; the old women had twigs put through the bored hole in their noses.

The men came back with some pine saplings; two of these they laid at the bottom of the grave, which was about five feet deep. On these pines they spread strips of bark, then a thick bed of Dheal twigs; then a woman handed a bag containing the belongings of the dead woman--boogurr they were called--to the oldest male relative, who was standing in the grave; he placed it as a pillow at one end. Then Hippi and the daughter's husband took each an end of the coffin and lowered it into the grave; the daughter cried loudly as they did so. Over the coffin they laid a rug, and on the rug they placed Beemunny's yam stick. Hippi signalled to the daughter, who then came with the other women close to the edge of the grave. She sat at one end, looked over into the grave, and called out: 'My mother! Oh, my mother! Come back to me, my mother! My mother that I have been with always, why did you leave me?' Then she wailed the death-wait, which the other women caught up. As the wail died away, Hippi said:

> 'She has gone from us; never as she was will she return.
> Never more as she once did will she chop honey.
> Never more with her gunnai dig yams.
> She has gone from us; never as she was to return.'

As he finished all the women wailed again, and loudest of all the daughter. Then the old man in the grave said:

> 'Mussels there are in the creek and plenty,
> But she who lies here will dig no more.
> We shall fish as of old for cod-fish,
> But she who lies here will beg no more oil,
> Oil for her hair, she will want no more.'

Then again the women wailed.

Old Hippi said, as the other man, in a sort of recitative

> 'Never again will she use a fire.
> Where she goes fires are not.
> For she goes to the women, the dead women,

And women can make no fires.
Fruit is there in plenty and grass seed,
But no birds nor beasts in the heaven of woman.'

Again the women wailed, wail after wail. Then they handed the remaining twigs of Dheal to the men, who laid them on the top of the coffin, then bark again over the twigs, and pine saplings on them, on top some old rugs.

While this was being done the old, old gins danced slowly a corroboree step round the edge of the grave, crooning a Goohnai-wurrai or dirge.

Then the men began to throw in the earth, the oldest male relative of the deceased standing in the grave to guard the body until the earth covered the coffin. As thud after thud went the earth in, the daughter shrieked and swayed over as if to fall into the grave, but her friend drew her back. She called 'Mother! mother!' took a sharp stone which was beside her and hit it against her head until the blood gushed out. They took the stone from her. There she sat rocking her body to and fro, wailing all the time, the other women wailing too, until the grave was quite covered in.

When it was filled in Hippi made another big smoke, thoroughly smoked himself, calling to all the men to do the same.

An old woman made a big smoke behind where the women were sitting; she called them one by one and made them stand in the thick of it for a while.

Hippi said something to her. I caught the word 'Innerah'--they called me Innerah, which meant literally a woman with a camp of her own. The old woman gave the smoke fire a stir, and out at once came a thick column of smoke circling round my guest and myself.

They covered the grave with logs and boughs and then swept round it.

All was over, we turned homewards. As we did so a flock of screeching gilahs flew over, their bright rose colouring lighting up the sombre scene where the only colour was that of the dark pines silhouetted against a sky from which the blue had now faded. Going home Bootha told me that the smoking process was to keep the spirits away, and to disinfect us from any disease the dead might have; and she said had we not been smoked the spirits might have followed us back to the house.

They would at once change their camp; the old one would be gummarl--a tabooed place; but before they left it they would burn smoke fires there to scare away the spirits.

I asked her why they swept round the grave. She said, in case the dead person had been poisoned or killed by magic; and, indeed, so little do they allow the possibility of death from natural causes, they even said old Beemunny had been given poison in her honey by an old-time rejected lover. Well, by sweeping round the grave they would see what track was on the swept place next morning, and according to that they would know to what totem the murderer belonged. If the track should be an iguana's, then one of the Beewee, or iguana totem, was guilty; if an emu, then one of the Dinewan, or emu totem, and so on.

Old Hippi joined me a little further on. He explained that the service was not as it would have been some years ago. That I knew, because when I first went to the station I had seen them going to funerals all decorated as if for corroborees. Round their waists, wrists, knees and ankles had been twigs of Dheal, the sacred tree, and the rest of their bodies had been painted.

Hippi said a great deal more would have been spoken and sung at the grave if the dead person had been a man. His spirit would have in a short sort of prayer been commended to Byamee, who would have been intreated to let the dead enter Bullimah (heaven), as he had kept the Boorah laws-that is, of course, if he had been initiated: the spirits of the uninitiated wander until they are reincarnated, and never enter Bullimah. One curious coincidence occurred in connection with this burial.

Seeing the droughty desolation of the country, as we walked to the grave, I asked old Bootha when she thought it would rain again. Coming very close to me she half whispered:

'In three days I think it; old woman dead tell me when she dying that "'sposin" she can send 'em rain, she send 'im three days when her Yowee bulleerul--spirit breath--go long Oobi Oobi.'

Beemunny died on Wednesday night. On Saturday when we went to bed the skies were as cloudless as they had been for weeks. In the middle of the night we were awakened by the patter of rain-drops on the iron roof. All night it rained, and all the next day.

It is said that a dead person always sends rain within a week of his death to wash out his tracks on earth.

One little black girl told me she always felt sad when she saw thunderclouds, because she thought some dead person had sent them.

As a rule, there is a good deal more shedding of blood over a grave than I saw. This blood offering is said to please the dead, being a proof to them of the affection of the living. It is funeral etiquette to prepare yourself with a weapon with which to shed this blood, but likewise etiquette for a friend to intervene and stop your self-mutilation.

On emerging from the grave the spirit finds the spirits of his dead relations waiting to go with him to Oobi Oobi, that is, a sacred mountain whose top towers into the sky, nearly touching Bullimah. The new spirit recognises his relations at once; they had, many of them, been round the death-bed visible at the last to the dying, though not to any of the watchers with him, though these are said sometimes to hear the spirit voices.

The spirit from the grave carried with him the twigs of the sacred Dheal tree which were placed over and under his body; he follows his spirit relations, dropping these twigs as he goes along, leaving thus a trail that those who follow may see. At the top of Oobi Oobi he finds the spirits called Mooroobeaigunnil, whose business it is to bridge over the distance a spirit has to traverse between the top of the mountain and Bullimah, the great Byamee's sky-camp.

One of these Mooroobeaigunnil seizes him and hoists him on to his shoulders; then comes another and hoists the first; and so on, until the one holding the spirit can lift him into Bullimah. As the spirit is hoisted in, one of the Mooroobeaigunnil, knocks the lowest one in the ladder of spirits down; thud to the earth come the rest, making a sound like a thunderclap, which the far away tribes hear, and hearing say:

'A spirit has entered Bullimah.'

Should a big meteor fall followed by a thunderclap, it is a sign that a great man has died. Should a number of stars shoot off from a falling star, it is a sign that a man has died leaving a large family. When a star is seen falling in the day-time, it is a sign that one of the Noongahburrah tribe dies.

In the olden time some of the tribes would keep a body at least five days. Then they would rub the outside black skin off, make an opening in the side of the body, take out the internal parts, fill it up with Dheal leaves. They would place the rubbed-off skin and internals in bark and put it in hollow trees. They would then bury the body, which they said would come up white.

Sometimes they would keep their dead for weeks, that they might easily extract the small joint bones with which to make poison.

A baby's body they would sometimes carry for years before burying, but it would usually have been well smoke-dried first, though not, I believe, invariably so.

Sometimes a body was kept so that relations from a distance might come and see for themselves the death was not the result of foul play.

After the body was filled up with Dheal leaves it was put into its bark coffin and smoke fires made round it.

As each relation arrived he was blindfolded and led up to the corpse, which was held up standing by some of the men. When the blindfolded relation came near, the bandage was taken off him and before him he saw standing his relation, whom he examined to see if wounds were visible. If signs of violence were apparent, the murderer had to be discovered and stand his trial.

He was given a shield to defend himself with. Every man had a right to throw a weapon at him; should he manage to defend himself successfully, as far as that crime was concerned he would be henceforth a free man, no stigma attaching to him whatever. In which, I fancy, the blacks show themselves a larger-minded people than their white supplanters, who make this world no place for repentance for wrong-doers, 'though they seek it with tears.' In the world's opinion there is no limit to a man's sentence. We read the letter of the Gospel, and leave the spirit of it to the blacks to apply.

Should there be a difficulty as to discovering the criminal, all the men of the tribes amongst whom the murderer could be stand round the coffin. A head man says to the corpse, 'Did such and such a man harm you?' naming, one after another, all the men. At the guilty one's name the corpse is said to knock a sort of rap, rap, rap.

That man has to stand his trial.

But as a rule the blacks like to bury their dead quickly, because the spirit haunts their neighbourhood or its late camp until the body is buried. Mysterious lights are said to be seen at night, and there is a general scare in camp-land until a corpse is safely buried.

There are variations in the funeral rites of nearly every tribe. Even in our district the dead were sometimes placed in hollow trees. I know of skeletons in trees on the edge of the ridge on which the home station was built. These are said to be for the most part the bodies of worthless women or babies.

In the coastal districts there are platforms in trees on which dead bodies were laid. In some places corpses are tied up in a sitting posture. The tying, they say, is to keep them secure when spirits come about, or body-snatchers for poison bones.

In some places the graves are covered with a sort of emu egg-shaped and sized lumps of copi; and also, when a widow's term of mourning was over, she would take the widow's cap-which was a sort of copi or gypsum covering put on wet to her head-and place it on the grave of her husband.

On the Narran the widows plaster their heads with copi or bidyi, as they call it, but so thinly that it cakes off. They renew it, and keep their heads covered with it for the allotted term of mourning, then just let it gradually all wear off.

Those widows' caps, having the imprint of nets inside them, are very old; for hair nets have been out of fashion for very many years in camp-land, so such rank as antique curios.

I don't think the small girl who thought when she grew up she'd choose to be a widow, would have thought so if she had been born black.

When a black woman's husband dies she has to cover herself with mud, and sleep beside a smouldering smoke all night. Three days afterwards, black fellows go and make a fire by the creek. They chase the widow and her sisters, who might have been her husband's wives, down to the creek. The widow catches hold of the smoking bush, puts it under her arm, and jumps into the middle of the creek; as the smoking bush is going out she drinks some of the smoky water. Then out she comes, is smoked at the fire; she then calls to those in the camp, and looks towards her husband's grave and calls again; his spirit answers, and the blacks call to her that they have heard him.

After that she is allowed to speak; she had been doomed to silence since his death, but for lamentations. She goes to the new camp, where another big smoke is made. She puts on her widow's cap, which, as it wears out, has to be renewed for many months; for some months, too, she keeps her face daubed with white.

Every time a stranger comes to the camp the widow has to make a smoke and smoke the camp again. The nearest of kin to her husband has a right to claim her as wife when her mourning is over.

Should a woman be left a widow two or three times there are sinister whisperings about her. She is spoken of as having a 'white heart'; and no man can live long, they say, with a woman having a 'white heart.'

The graves in some parts of Australia are marked by carved trees; only a few painted upright posts marked them on the Narran.

A tabooed camp has always a marked tree-just a piece of bark cut off and some red markings made on the wood, which indicate that the place is gummarl.

Any possessions of the dead not buried with them are burnt, except the sacred stones; they are left to the wirreenun nearest of kin to the dead person.

Lately a case came under my notice of the taboo extended to the possessions of dead people.

A black man having two horses died. Neither his widow nor her mother would use those horses, even when he had been dead over a year. They would walk ten or twelve miles for their rations and carry them back, rather than use those horses before the term of mourning was over.

The widow was one of my particular friends, but she would not come to see me because her husband had been at the house shortly before he died. She camped nearly a mile away, and I went to see her there. After he had been dead about a year, she came to see me; but before she did so her mother walked all round the out-buildings, garden, yards, etc., with a bunch of smoking Budtha, crooning little spirit songs.

SOMETHING ABOUT STARS AND LEGENDS

VENUS in the Summer evenings is a striking object in the western sky. Our Venus they call the Laughing Star, who is a man. He once said something very improper, and has been laughing at his joke ever since. As he scintillates you seem to see him grinning still at his Rabelais-like witticism, seeing which the aborigines say:

'He's a rude old man, that Laughing Star.'

The Milky Way is a warrambool, or water overflow; the stars are the fires, and the dusky haze the smoke from them, which spirits of the dead have lit on their journey across the sky. In their fires they are cooking the mussels they gather where they camp.

There is one old man up there who was once a great rainmaker, and when you see that he has turned round as the position of the Milky Way is altered, you may expect rain; he never moves except to make it.

A waving dark shadow that you will see along the same course is Kurreah, the crocodile.

To get to the Warrambool, the Wurrawilberoo, two dark spots in Scorpio, have to be passed. They are devils who try to catch the spirits of the dead; sometimes even coming to earth, when they animate whirlwinds and strike terror into the blacks. The old men try to keep them from racing through the camp by throwing their spears and boomerangs at them.

The Pleiades are seven sisters, as usual, the dimmed ones having been dulled because on earth Wurrunnah seized them and tried to melt the crystal off them at a fire; for, beautiful as they were with their long hair, they were ice-maidens. But he was unsuccessful beyond dulling their brightness, for the ice as it melted put out the fire. The two ice-maidens were miserable on earth with him, and eventually escaped by the aid of one of their 'multiplex totems,' the pine-tree. Wurrunnah had told them to get him pine bark. Now the Meamei--Pleiades--belong to the Beewee totem, so does the pine-tree. They chopped the pine bark, and as they did so the tree telescoped itself to the sky where the five other Meamei were, whom they now joined, and with whom they have remained ever since. But they who were polluted by their enforced residence with the earth-man never shone again with the brightness of their sisters. This legend was told emphasising the beauty of chastity.

Men had desired all the sisters when once they travelled on earth, but they kept themselves unspotted from the world, with the exception of the two Wurrunnah captured by stratagem.

Orion's Sword and Belt are the Berai-Berai--the boys--who best of all loved the Meamei, for whom they used to hunt, bringing their offerings to them; but the ice-maidens were obdurate and cold, disdaining lovers, as might be expected from their parentage. Their father was a rocky mountain, their mother an icy mountain stream. But when they were translated to the sky the Berai-Berai were inconsolable. They would not hunt, they would not eat, they pined away and died. The spirits pitied them and placed them in the sky within sound of the singing of the Meamei, and there they are happy. By day they hunt, and at night light their corroboree fires, and dance to the singing in the distance. just to remind the earth-people of them, the Meamei drop down some ice in the winter, and they it is who make the winter thunderstorms.

Castor and Pollux, in some tribes, are two hunters of long ago.

Canopus is Womba, the Mad Star, the wonderful Weedah of long ago, who, on losing his loves, went mad, and was sent to the sky that they might not reach him; but they followed, and are travelling after him to this day, and after them the wizard Beereeun, their evil genius, who made the mirage on the plains in order to deceive them, that they and Weedah might be lured on by it and perish of thirst.

When they escaped him Beereeun threw a barbed spear into the sky, and hooked one spear on to another until he made a ladder up which he climbed after them; and across the sky he is still pursuing them.

The Clouds of Magellan are the Bralgah, or Native Companions, mother and daughter, whom the Wurrawilberoo chased in order to kill and eat the mother and keep the daughter, who was the great dancer of the tribes. They almost caught her, but her tribe pursued them too quickly; when, determined that if they lost her so should her people, they chanted an incantation and changed her from Bralgah, the dancing-girl, to Bralgah, the dancing-bird, then left her to wander about the plains. They translated themselves on beefwood trees into the sky, and there they are still.

Gowargay, the featherless emu, is a debbil-debbil of water-holes; he drags people who bathe in his holes down and drowns them, but goes every night to his sky-camp, the Coalpit, a dark place by the Southern Cross, and there he crouches. Our Corvus, the crow, is the kangaroo.

The Southern Crown is Mullyan, the eagle-hawk. The Southern Cross was the first Minggah, or spirit tree a huge Yaraan, which was the medium for the translation of the first man who died on earth to the sky. The white cockatoos which used to roost in this tree when they saw it moving skywards followed it, and are following it still as Mouyi, the pointers. The other Yaraan trees wailed for the sadness that death brought into the world, weeping tears of blood. The red gum which crystallises down their trunks is the tears.

Some tribes say it was by a woman's fault that death came into the world.

This legend avers that at first the tribes were meant to live for ever. The women were told never to go near a certain hollow tree. The bees made a nest in this tree; the women coveted the honey, but the men forbade them to go near it. But at last one woman determined to get that honey; chop went her toma-hawk into that hollow trunk, and out flew a huge bat. This was the spirit of death which was now let free to roam the world, claiming all it could touch with its wings.

Of eclipses there are various accounts. Some say it is Yhi, the sun, the wanton woman, who has overtaken at last her enemy the moon, who scorned her love, and whom now she tries to kill, but the spirits intervene, dreading a return to a dark world. Some say the enemies have managed to get evil spirits into each other which are destroying them. The wirreenuns chant incantations to oust these spirits of evil, and when the eclipse is over claim a triumph of their magic.

Another account says that Yhi, the sun, after many lovers, tried to ensnare Bahloo, the moon; but he would have none of her, and so she chases him across the sky, telling the spirits who stand round the sky holding it up, that if they let him escape past them to earth, she will throw down the spirit who sits in the sky holding the ends of the Kurrajong ropes which they guard at the other end, and if that spirit falls the earth will be hurled down into everlasting darkness.

So poor Bahloo, when he wants to get to earth and go on with the creation of baby girls, has to sneak down as an emu past the spirits, hurrying off as soon as the sun sinks down too.

Bahloo is a very important personage in legends.

When the blacks see a halo round the moon they say,

'Hullo! Going to be rain. Bahloo building a house to keep himself dry.'

All sorts of scraps of folk-lore used to crop out from the little girls I took from the camp into the house to domesticate. When storms were threatening, some of the clouds have a netted sort of look, something like a mackerel sky, only with a dusky green tinge, they would say: 'See the old man with the net on his back; he's going to drop some hailstones.'

Meteors always mean death; should a trail follow them, the dead person has left a large family.

Comets are a spirit of evil supposed to drink up the rain-clouds, so causing a drought; their tails being huge families all thirsty, so thirsty that they draw the river up into the clouds.

Every natural feature in any way pronounced has a mythical reason for its existence, every peculiarity in bird life, every peculiarity in the trees and stones. Besides there are many mythical bogies still at large, according to native lore, making the bush a gnome-land.

Even the winds carry a legend in their breath.

You hear people say they could have 'burst with rage,' but it is left to a black's legend to tell of a whole tribe bursting with rage, and so originating the winds.

There was once an invisible tribe called Mayrah. These people, men and women, though they talked and hunted with them, could never be seen by the other tribes, to whom were only visible their accoutrements for hunting. They would hear a woman's voice speak to them, see perhaps a goolay in mid-air and hear from it an invisible baby's cry; they would know then a Mayrah woman was there. Or a man would speak to them. Looking up they would see a belt with weapons in it, a forehead band too, perhaps, but no waist nor forehead, a water-vessel invisibly held: a man was there, an invisible Mayrah. One of these Mayrah men chummed with one of the Doolungaiyah tribe; he was a splendid mate, a great hunter, and all that was desirable, but for his invisibility. The Doolungaiyah longed to see him, and began to worry him on the subject until at last the Mayrah became enraged, went to his tribe, and told them of the curiosity of the other tribes as to their bodily forms. The others became as furious as he was; they all burst with rage and rushed away roaring in six different directions, and ever since have only returned as formless wind to be heard but never seen. So savagely the Mayrah howled round the Doolungaiyah's camp that he burrowed into the sand to escape, and his tribe have burrowed ever since.

Three of the winds are masculine and three feminine. The Crow, according to legend, controls Gheeger Gheeger, and keeps her in a hollow log. The Eagle-hawk owns Gooroongoodilbaydilbay, and flies with her in the shape of high clouds. Yarragerh is a man, and he has for wives the Budtha, Bibbil, and Bumble trees, and when he breathes on them they burst into new shoots, buds, flowers, and fruits, telling the world that their lover Yarragerh, the spring, has come.

Douran Doura woos the Coolabah, and Kurrajong, who flower after the hot north wind has kissed them.

The women winds have no power to make trees fruitful. They can but moan through them, or tear them in rage for the lovers they have stolen, whom they can only meet twice a year at the great corroboree of the winds, when they all come together, heard but never seen; for Mayrah, the winds, are invisible, as were the Mayrah, the tribe who in bursting gave them birth.

Yarragerh and Douran Doura are the most honoured winds as being the surest rain-bringers. In some of the blacks' songs Mayrah is sung of as the mother of Yarragerh, the spring, or as a woman kissed into life by Yarragerh putting such warmth into her that she blows the winter away. But these are poetical licences, for Yarragerh is ordinarily a man who woos the trees as a spring wind until the flowers are born and the fruit formed, then back he goes to the heaven whence he came.

Then there are the historical landmarks: Byamee's tracks in stone, and so on, and the battle-fields, too, of old tribal fights. Just in front of our station store was a gnarled old Coolabah tree covered with warty excrescences, which are supposed to be seats for spirits, so showing a spirit haunt.

In this particular tree are the spirits of the Moungun, or armless women, and when the wind blows you could hear them wailing. Their cruel husband chopped their arms off because they could not get him the honey he wanted, and their spirits have wailed ever since.

Across the creek is another very old tree, having one hollow part in which is said to be secreted a shell which old Wurrunnah, the traveller of the tribes, and the first to see the sea, brought back. No one would dare to touch the shell. The tribe of a neighbouring creek, when we were first at the station, used to threaten to come and get it, but the men of the local tribe used to muster to protect it from desecration even at the expense of their lives.

The Minggah by the garden I have told you of before. Further down the creek are others.

At Weetalibah was the tree from which Byamee cut the first Gayandi. This tree was burnt by travellers a few years ago. The blacks were furious: the sacred tree of Byamee burnt by the white devils! There are trees, too, considered sacred, from which Byamee cut honey and marked them for his own, just as a man even now, on finding a bee's nest and not being able to stay and get it, marks a tree, which for any one else to touch is theft.

A little way from the head station was an outcrop of white stones. These are said to be fossilised bones of Boogoodoogah-dah's victims. She was a cannibal woman who had hundreds of dogs; with them she used to round up blacks and kill them, and she and her dogs ate them. At last she was outwitted and killed herself, and her spirit flew out as a bird from her heart. This bird haunts burial grounds, and if in a drought any one can run it down and make it cry out, rain will fall.

During a drought one of these birds came into my garden, hearing which the blacks said rain would come soon, and it did. In another drought when the rainmakers had failed, some of the old blacks saw a rain-bird and hunted it, but could not get it to call out.

Geologists say there should be diamonds along some of the old water-courses of the Moorilla ridges. Perhaps the white stone that the blacks talk about, which shows a light at night, and has, they say, a devil in it, is a diamond. Ruskin rather thought there was a devil in diamonds, making women do all sorts of evil to possess them. The blacks told me that a Queensland tribe had a marvellous stone which at great gatherings they show. Taking those who are privileged to see it into the dark, there they suddenly produce it, and it glows like a star, though when looked closely at in daylight seems only like a large drop of rain solidified. This stone, they said, has to be well guarded, as it has the power of self-movement, or rather, the devil in it can move it.

The greatest of local landmarks is at Brewarrina; this is the work of Byamee and his giant sons, the stone fisheries made in the bed of the Barwon.

At Boogira, on the Narran Lake, is an imprint in stone of Byamee's hand and foot, which shows that in those days were giants. There it was that Byamee brought to bay the crocodiles who had swallowed his wives, from which he recovered them and restored them to life.

At Mildool is a scooped-out rock which Byamee made to catch and hold water; beside it he hollowed out a smaller stone, that his dog might have a drinking-place too. This recurrence of the mention of dogs in the legends touching Byamee looks as if blacks at all events believed dogs to have been in Australia as long as men.

At Dooyanweenia are two rocks where Byamee and Birrahgnooloo rested, and to these rocks are still sticking the hairs he pulled from his beard, after rubbing his face with gum to make them come out easily.

At Guddee, a spring in the Brewarrina district, every now and then come up huge bones of animals now extinct. Legends say that these bones are the remains of the victims of Mullyan, the eagle-hawk, whose camp was in the tree at the foot of which was the spring. This tree was a tree of trees; first, a widely spreading gum, then another kind, next a pine, and lastly a midgee, in which was Mullyan's camp, out of which the relations of his victims burnt him and his wives, and they now form the Northern Crown constellation. The roots of this gigantic tree travelled for miles, forming underground watercourses. At Eurahbah and elsewhere are hollowed-out caves like stones; in these places Birrahgnooloo slept, and near them, before the stock trampled them out, were always to be found springs made at her instigation for her refreshment; she is the patroness of water.

At Toulby and elsewhere are mud springs. It is said that long ago there were no springs there, nor in the Warrego district, and in the droughts the water-courses all dried up and the blacks perished in hundreds. Time, after time this happened, until at last it seemed as if the tribes would be exterminated. The Yanta--spirits--saw what was happening and felt grieved, so they determined to come and live on the earth again to try and bring relief to the drought-stricken people. Down they came and set to work to excavate springs. They scooped out earth and dug, deeper and deeper, until at length after many of them gave in from exhaustion, those that were left were rewarded by seeing springs bubble up.

The first of those that they made was at Yantabulla, which bears their name to this day.

The blacks were delighted at having watering-places which neither a drought nor the fiercest sun could dry up. The Yantas were not contented with this nor with the other springs they made. They determined to excavate a whole plain, and turn it into a lake so deep that the sun could never dry it, and which would be full of fish for the tribes.

They went to Kinggle and there began their work. On they toiled unceasingly, but work as they would they could not complete their scheme, for one after another wearied and died, until at last nothing was left on the plain but the mud springs under the surface and the graves of the Yantas on top. No blacks will cross Kinggle plains lest some of these spirits arise through the openings of their graves.

This legend shows what a disheartening country the West is in a drought. When even the spirits gave in, how can ordinary men succeed? But indeed it is not ordinary men who do, but our 'Western heroes,' as Will Ogilvie calls them, who wear their cross of bronze on neck and cheek in the country where 'the green fades into grey.'

THE TRAPPING OF GAME

SOME of the blacks' methods of catching game I have seen practised, some have long since died out of use.

Of course the sportsmen knew the favourite watering-holes of the game. At such a place they made a rough break at each side, leaving an opening where the track was. Along this track they would lay a net with one end on the edge of the water; in the water they put sticks on the ends of which the birds rest to drink, the other ends are out in the trap. They would make a hole low down on each side of the net, and a man would hide in each.

A bird's watering-place, where the blacks trap them, is called Dheelgoolee. When the Dheelgoolee trapping begins, on the first day those who go out hunting must bring home their game alive to give the man at the Dheelgoolee luck. Then they never try to catch an emu or kangaroo, only iguana, opossum, piggiebillah, paddy melon, or bandicoot, all of which could be brought home alive. But after the first day they can kill as they go along.

All day some birds come to the Dheelgoolee-pigeons, gilahs, young crows, and others, and the man watching catches them. When the game was thick on the net, the men in the holes would catch hold of the ends of the sticks in the net and quickly turn them over the lower ends, thus entrapping all on the net. In the evening turkeys and such things as water at night-time, amongst which are opossums and paddy melons, would be trapped.

Ducks were trapped, too, by making bough breaks across the shallow part of the creek, with a net across the deep part from break to break. A couple of the men would go up stream to hunt the ducks down, and some would stay each side of the net armed with pieces of bark. The two hunters up stream frightened the ducks off the water, and sent them flying down stream to the trap. Should they seem flying too high as if to pass, the blacks would throw the pieces of bark high in the air, imitating, as they did so, the cry of hawks. Down the ducks would fly turning back; some of the men would whistle like ducks, others would throw bark again, giving the hawk's cry, which would frighten the birds, making them double back into the net, where they were quickly despatched by those waiting.

Murrahgul is another trap. This is a yard made all round a waterhole with one opening; about this opening they will fasten, from stumps or logs, strong strings with a slipping knot. The game, emu or kangaroo, would probably step into one of these string nooses, would try to pull its leg out; the harder it pulled the tighter the knot. Or the blacks might have put a sort of cross-bar overhead at the entrance, with hanging strings having a slip knot; in would go an emu's head, the bird would rush on and be strangled.

Boobeen is a primitive cornet, a hollowed piece of Bibbil wood, one end partially filled up with pine gum, and ornamented outside with carvings. To blow through it is an art, and the result rather like a big horn. The noise is said to be very like an emu's cry, and this emu bugle will certainly, they say, draw towards it a gundooee, or solitary emu.

The blacks used on the sandhills to make a deep hole to hide themselves in, usually only one though. From this hole they would run out a drain for about thirty yards. The man with the Boobeen would have a little break of bushes round him;

scattered over the leaves he'd have emu feathers, and then he would have a strong string, on the end of which he would have a small branch with this he would place about midway emu feathers on it; down the drain.

When the emu answers the Boobeen's call, the bugler gets lower and slower with his call. The emu sees the feathered thing in the drain, comes inquisitively up and sniffs at it. The man in the hole pulls in the string slowly; the emu follows, on, on, until heedlessly he steps on a Murrahgul, or string trap, and is caught. The hunters would sometimes stalk kangaroo, holding in front of them boughs of trees or bushy young saplings, closing silently in and in, until at last the kangaroo were so closely surrounded by men armed with boondees and spears that there was no escape for them.

For catching emu they had a net made of string as thick as a clothes-line. These nets were made either of Kurrajong (Noongah) bark, or of Burraungah grass. The Kurrajong bark is stripped off the trees, beaten, chewed, and then teased. Then it was taken and rubbed, principally by the women on their legs, into strands.

The grass was used preferably to Kurrajong bark, as it was easier to work. The process of preparation was as follows:--

A hole was dug in the ground, some fire put in it, a. quantity of ordinary grass was put on the top of the coals, and on top of that a heap of Burraungah grass, that topped with ordinary grass.

Water was sprinkled over it all and the hole earthed up.

When it had been in long enough the earth was cleared away, and the grass, which was quite soft, taken out. It was then chewed and worked like the Kurrajong bark, than which it was much more pliable.

String was made of various thicknesses according to what it was required for.

Fishing nets were always smoked before being used, and all nets had little charm songs sung over them. In netting, their only implement was a piece of wood to wind their string on. An emu net was about five feet high, and between two and three hundred yards long.

When any one discovered a setting emu, they used not to disturb her at once and get her eggs, but returned to the camp, singing as they neared it a song known as the Noorunglely, or setting emu song; those in camp would recognise it, and sing back the reply. The black fellows having learnt where the nest was, would get their net and go out to camp near it. All that evening they would have an emu-hunting corroboree. The next morning at daylight they would erect their net into a sort of triangular-shaped yard, one side open. Black fellows would be stationed at each end of the net, and at stated intervals along the mirroon, as the net was called. When the others were all ready some of the blacks would make a wide circle round the emu, leaving open the side towards the net; they would close in gradually until they frightened the emu off her nest; she would run in the direction where she saw no black fellows and where the net was; the black fellows closing in behind, followed quickly. Poor Noorunglely floundered into the net, up rushed a black fellow and, seizing her, wrung her neck. Having secured her, they would next secure her eggs; that they might be a trifle stale was a matter of indifference to them.

Another old method was by making sort of brush yards and catching the emus in these.

One modern way is to run them down with kangaroo dogs, the same way with kangaroo; but at one time still another method obtained. A black fellow would get a long spear and fasten on the end a bunch of emu feathers. When he sighted an emu he would climb a tree, break some boughs to place beneath him, if the trees were thinly foliaged, to hide him from the emu, then he would let his spear dangle down. The emu, a most inquisitive bird, seeing the emu feathers, would investigate. Directly the bird was underneath the tree, the black fellow would grip his spear tightly and throw it at the emu, rarely, if ever, failing to hit it, though the emu might run wounded for a short distance, but the black fellow would be quickly after it to give it happy despatch.

If the emu got a good start even, it was easily tracked by the trail of blood. It has happened that a black fellow has not found his emu until the next day, when it was dead and the spear still in it; but usually very soon after the wounded birds start running the spear is shaken out.

Sometimes the blacks killed birds with their boomerangs, ducks in particular. I fancy this killing of ducks by a well-thrown boomerang is one of the feats that black fellows allow themselves to blow about. Every man has usually one subject, a speciality he considers of his own, and on that subject he waxes eloquent.

Pigeons, gilahs, and plains turkeys are also killed with boomerangs. Blacks' fishing-nets are about ten feet by five, a stick run through each end, for choice of Eurah wood. Eurah is a pretty drooping shrub with bell-shaped spotted flowers, having a horrible smell. The wood is very pliable. It is sometimes used instead of the sacred Dheal at funerals.

Two of the fishermen take the net into the creek, one at each end; they stand in a rather shallow place, holding the net upright in the water. Some other blacks go up stream and splash about, frightening the fish down towards the net. When those holding the net feel the fish in it, they fold the two sticks together and bring the net out.

To catch fish they also make small weirs and dams of stones, with narrow passages of stones leading to them. The fish are swept by the current into these yards, and there either caught by the blacks with their hands, or speared. The most celebrated of these stone fish-traps is at Brewarrina on the Barwon. It is said to have been made by Byamee, the god and culture-hero of these people, and his giant sons. He it was who established the rule that there should be a camping-ground in common for the various tribes where, during the fishing festival, peace should be strictly kept, all meeting to enjoy the fish, to do their share towards preserving the fisheries.

Each tribe has its particular yards; for another to take fish from these is theft. Each tribe keeps its yards in repair, replacing stones removed by floods, and so on.

These stony fish mazes are fully two hundred yards in length, substantially built; some huge boulders are amongst the stones which form these most intricate labyrinthine fish yards, which as traps are eminently successful, many thousands of Murray cod and other fish being caught in them.

Dingo pups, in the days when dingoes were plentiful, were a most esteemed delicacy. To eat dog is dangerous for a woman, as causing increased birth-pangs; that suggests dog must be rather good eating, some epicure wirreenun scaring women off it by making that assertion.

Ant larvæ, a special gift from some spirit in the stars, and frogs, are also thought good by camp epicures.

The blacks smear themselves over with the fat of fish or of almost any game they catch. It is supposed to keep their limbs supple, and give the admired ebony gloss to their skins which, by the way, are very fine grained. After a flood, when the water is running out of the tributaries of the creek, the blacks make a bough break beginning on each bank and almost meeting in the middle; across the gap they place a fishing-net which folds in like a bag, thus forming a fish-trap in which are caught any number of fish. Crayfish and mussels they caught by digging down their holes in the mud for them. Their mode of catching shrimps was very (with all apologies to scientists for using the word) primitive. Quite nude, the women sit down in the water, let the shrimps bite them; as they nip, seize them.

Iguanas burrow into the soft sand ridges and there remain during the winter, only coming out after the Curreequinquins-- butcher birds--one of their sub-totems, sing their loudest to warn them that the winter is gone, calling Dooloomai, the thunder, to their aid lest their singing is not heard by their relations, who after the storms come out again in as good condition as when they disappeared.

Black men do not approve of women cooks. At least the old men, under the iron rule of ancient custom, will not eat bread made by gins, nor would they eat iguana, fish, piggiebillah, or anything like that if the inside were removed by a woman, though after themselves having prepared such things, they allow the gins to cook them--that is, if they have not young children or are enceinte; under those conditions they are unclean.

FORAGING AND COOKING

IT is very strange to me to hear the average white person speak of the blacks collectively as having no individuality, for really they are as diverse in characteristics as possible; no two girls I had in the house but were totally different.

There has been too much generalisation about the blacks. For instance, you hear some people assert all blacks are trackers and good bushmen. That there are some whose tracking power is marvellous is true, but they are not the rule, and a black fellow off his own beat is often useless as a bushman.

So with their eyesight; what they have been trained to look out for they see in a marvellously quick way, or so it seems to us who have not in their lines the same aptitude. Of course, for seeing things at a distance a black has the advantage, unless the white has had the same open-air life. Some white bushmen are as good as any blacks.

Nimmaylee, a little black girl who lived in the house, used to tell me all sorts of bush wonders, as we went in the early summer mornings for a swim in the river. She was a great water-baby, with rather a contempt for my aquatic limitations. Then she thought it too idiotic to want to dry yourself with a towel,--just like a mad white woman!

White people were an immense joke to Nimmaylee. She conformed to their rules as one playing a new game. She has a little brother as black as herself. She has a substantial pair of legs, but his are so thin and his little body so round that he looks like a little black spider.

Nimmaylee is quite an authority on corroborees, knowing ever so many different steps, from the serpentile trail of the codfish to the mimic fight. The songs she knows too. She used, when she lived in the camp, to marshal in a little crowd of camp children, and put them through a varied performance for my benefit.

These performances were of daily occurrence when the fruit was ripe, for Nimmaylee's capacity for water-melon was practically unlimited.

Nimmaylee was a wonderful little fisherwoman; she delighted in a fishing expedition with me. Off we used to go with our lines, worms or frogs for bait, or perhaps shrimps or mussels if we were after cod. If we were successful, Nimmaylee would string the fish on a stick in a most professional manner, and carry them with an air of pride to the cook. She attributes her fishing successes to a charm having been sung over her to that end as a baby.

Accompanied by some reliable old 'gins' and ever so many piccaninnies, I used to take long walks through the 'bush.'

How interesting those blacks made my bush walks for me! Every ridge, plain, and bend had its name and probably legend; each bird a past, every excrescence of nature a reason for its being.

Those walks certainly at least modified my conceit. I was always the dunce of the party--the smallest child knew more of woodcraft than I did, and had something to tell of everything. Seeing Oogahnahbayah, a small eagle-hawk, flying over, they would say, 'He eats the emu eggs.' He flies over where the emu is sitting on her eggs and makes a noise hoping to frighten the bird off; having done so, he will drop a stone on the eggs. If the emu is not startled off the nest, the hawk will fly on, alight at some distance, and walk up like a black fellow, still with the stone in his beak, to the nest; off the emu will go, then the hawk bangs the eggs with the stone until he breaks them. He throws the stone on one side, has a feed of emu eggs, and goes off, leaving poor Moorunglely, the sitting emu, to come back and find her eggs all destroyed. As the narrative ended, the little would look quite sad, and say 'Nurragah!' 'Poor thing!' at the thought of the domestic tragedy in bird life.

I had to hear the stingless little native bees humming before I could see them; and as to knowing which tree had honey in it, unless I saw the bees, that was quite beyond me, while a mere toddler would point triumphantly to a 'sugar-bag' tree, recognising it as such by the wax on its fork, black before rain, yellowish afterwards.

This honey is good strained, but as the blacks get it, it is all mixed up with dirty wax and dead bees.

I deplored the sacrifice of the bees one day, but was told it was all right. Whoever had chopped the nest out would take home the waxy stick they had used to help get the honey out; they would throw the stick in the fire, then all the dead bees would go to a paradise in the skies, whence next season they would send Yarragerh Mayrah, the Spring Wind, to blow the flowers open, and then down they would come to earth again. One year the manna just streamed down the Coolabah and Bibbil trees; it ran down like liquid honey, crystallising where it dropped.

The old blacks said, 'It is a drought now, but it will be worse. Byamee has sent the manna by the little Dulloorah birds and the black ants, because there will be no flowers for the bees to get honey from, so he has sent this manna.' Each time he has done so, a great drought has followed, and indeed it was followed by one of the worst droughts Australia has ever known. Byamee, it is said, first sent them the manna because their children were crying for honey, of which there was none except in the trees that Byamee, when on earth, had marked for his own. The women had murmured that they were not allowed to get this; but the men were firm, and would neither touch it nor let them touch it, which so pleased Byamee that he sent the manna, and said he always would when a long drought threatened.

A great chorus of 'My Jerhs' would tell something was sighted.

It might be the track of a piggiebillah porcupine. This track was followed to a hollow log; then came the difficulty, how to get it out, for porcupines cling tightly with their sharp claws, and all a dog can do where a piggiebillah is concerned is to bark, their spines are too much to tackle at close quarters. But the old gins are equal to the occasion: a tomahawk to chop the log, and a yam-stick to dislodge the porcupine, who takes a good deal of killing before he is vanquished.

They say a fully initiated man can sing a charm which will make a piggiebillah relax his grip and be taken captive without any trouble. The piggiebillahs burrow into the sand and leave their young there as soon as the faintest feel of a spine appears. The baby piggiebillahs look like little indiarubber toys.

The opossums all disappeared from our district. When we were first there they were very numerous and used to make raids at night to my rose-bushes-great havoc the result. It is said a very great wirreenun-wizard-willed them away so that his enemy, whose yunbeai, or personal totem, the opossum was, should die. This design was frustrated by counter magic; two powerful wizards appeared and, acting in concert, put a new yunbeai into the dying man; he recovered.

When the opossums were about the blacks used to see their scratched tracks on the trees, and chop or burn them out. They miss the opossums very much, for not only were they a prized food, but their skins made rugs, their hair was woven into cords of which were made amulets worn on the forearm or head against sickness, and with no modern instrument can they so well carve their weapons, as with an opossum tooth. Naturally their desire is to see Moodai, the opossum, return; to that end a wirreenun is now singing incantations to charm him back.

Opossum hunters had a way of bringing them home strung round their necks; very disagreeable, I should think, but custom, that tyrant, rules it so. The old gins dug out yams vigorously; some were eaten raw, others were kept for cooking.

To cook them they dug out a hole, made a fire in it, put some stones on the fire, then, when the stones were heated and the fire burnt down, they laid some leaves and grass on the stones, sprinkled some water, then put on the yams, on top of them more grass, sprinkled more water, then more grass and a. thick coating of earth, leaving the yams to cook.

Several other roots they cooked and ate. Raw they ate thistle tops, pigweed, and crowfoot, with great relish. Their game they cooked as follows. Kangaroo were first singed, cleaned out, and filled with hot stones, then put on the top of a burnt-down fire, hot ashes heaped all over them. The blacks like their meats with the gravy in, very distinctly red gravy. Emu were plucked, the insides taken out, and the birds filled up with hot stones, box leaves, and some of their own feathers. A fire was made in a hole; when it was burnt down, leaves and emu feathers were put in it, on top of these the bird, on top of it leaves and feathers again, then a good layer of hot ashes, and over all some earth.

The piggiebillahs were first smoked so that their quills might be easily knocked off. This done, the insides were taken out, then the piggiebillahs were put in little holes made beside the fire, and covered over with hot ashes, as were also opossums, ducks and other birds, iguanas and fish.

Ducks were plucked by our tribe, but in some places they were encased thickly in mud, buried in the ashes to cook, and, when done, the plaster of mud would be knocked off, and with it would come all the feathers.

The insides of iguanas and fish are taken out all in one piece. Each fish carries in its inside a representation of its Minggah-- spirit tree; by drying the inside and pressing it you can plainly see the imprint of the tree.

When we go bathing, the blacks tell me that the holes in the creek filled with gum leaves are codfish nests. They say too, that when they beat the river to drive the fish out towards the net waiting for them, that they hear the startled cod sing out.

Mussels and crayfish are cooked in the ashes.

The seagulls, which occasionally we used to see inland, are said to have brought the first mussels to the back creeks.

Emu eggs the blacks roll in hot ashes, shake, roll again; shake once more, and then bury them in the ashes, where they are left for about an hour until they are baked hard, when they are eaten with much relish and apparently no hurt to digestion, though one egg is by no means considered enough for a meal in spite of its being equal to several eggs of our domestic hen.

Not only are the blacks very particular in the way their game is carved or divided, but also in the distribution of the portions allotted to each person. The right to a particular part is an inherited one. No polite offering of a choice to an honoured guest, no suggestion of the leg or wing. You may loathe the leg of a bird as food, but at a black fellow's feast, if convention ordains that as your portion, have it you must; just as each rank in society had its invariable joint in early mediaeval Ireland.

The seeds of Noongah--a sterculia--and Dheal, were ground on their flat dayoorl-stones and made into cakes, which they baked, first on pieces of bark beside the fire to harden them, then in the ashes. These dayoorl, or grinding-stones, are handed down from generation to generation, being kept each in the family to whom it had first belonged. Should a member of any other use it without permission, a fight would ensue. Some of these stones are said to have spirits in them; those are self-moving, and at times have the power of speech. I have neither seen them move nor heard them speak, though I have a couple in my possession. I suppose the statement must be taken on faith; and as faith can move mountains, why not a dayoorl-stone?

The so-called improvident blacks actually used to have a harvest time, and a harvest home too. When the doonburr, or seed, was thick on the yarmmara, or barley-grass, the tribes gathered this grass in quantities.

First, they made a little space clear of everything, round which they made a brush-yard. Each fresh supply of yarmmara, as it was brought in by the harvesters, was put in this yard. When enough was gathered, the brush-yard was thrown on one side, and fire set to the grass, which was in full ear though yet green. While the fire was burning, the blacks kept turning the grass with sticks all the time to knock the seeds out. When this was done, and the fire burnt out, they gathered up the seed into a big opossum-skin rug, and carried it to the camp.

There, the next day, they made a round hole like a bucket, and a square hole close to it. These they filled with grass seed. One man trampled on the seed in the square hole to thresh it out with his feet; another man had a boonal, or stick, about a yard long, rounded at one end, and nearly a foot broad; with this he worked the grass in the round hole, and as he worked the husks flew away.

It took all one day to do this. The next day they took the large bark wirrees, canoe-shaped vessels, which when big like these are called yubbil. They put some grain in these, and shook it up; one end of the yubbils being held much higher than the other, thus all the dust and dirt sifted to one end, whence it was blown off. When the grain was sufficiently clean, it was put away in skin bags to be used as required, being then ground on the large flat dayoorl-stones, with a smaller flat stone held in both hands by the one grinding; this stone was rubbed up and down the dayoorl, grinding the seed on it, on which, from time to time, water was thrown to soften it.

When ground, the grain was made into little flat cakes, and cooked as the tree-seed cakes were. When the harvesting of the yarmmara was done, a great hunt took place, a big feast was prepared, and a big corroboree held night after night for some time.

The two principal drinks were gullendoorie--that is, water sweetened with honey; and another made of the collarene, or flowers of the Coolabah (grey-leaved box), or Bibbil (poplar-leaved box) flowers, soaked all night in binguies (canoe-shaped wooden vessels) of water. Just about Christmas time the collarene is at its best; and then, in the olden days, there were great feasts and corroborees held.

The flat dayoorl-stones on which the seeds are ground with the smaller stone, are like the 'saddle-stone querns' occasionally found in ancient British sites. These primitive appliances preceded the circular rotatory querns in evolution, and as the monuments prove were used in ancient Egypt. I cannot say whether, amongst the Euahlayi, there was a recognised licence as to exchange of wives on these festal occasions, or at boorahs. If the custom existed, I was not told of it by the blacks; but it is quite possible that, unless I made inquiries on the subject, I would not be told.

COSTUMES AND WEAPONS

HAVE seen a coloured king simply smirking with pride, in what he considered modern full dress--a short shirt and an old tall hat.

And I suppose, as far as actual clothing went, it was an advance on the old-time costume of paint and feathers. A black woman's needle was a little bone from the leg of an emu, pointed. Her thread was sinews of opossums, kangaroos, and emus; that was all that was necessary for her plain sewing, which was plain indeed.

Her fancy work consisted of netting dillee, goolays, or miniature hammocks to sling her baby across her back, or, failing a baby, her mixed possessions, from food to feathers; her larder and wardrobe in one.

Her costume being simple in the extreme did not require much room. It consisted of a goomillah, which was a string wound round the waist, made of opossum sinews, and in front, hanging down for about a foot, were twisted strands of opossum hair. A bone, or on state occasions a green twig, stuck through the cartilage of her nose, a string net over her hair, or perhaps only a fillet, or a kangaroo's tooth fastened to her front lock, gum balls dried on side-locks, an opossum's hair armlet, and perhaps a reed bead necklet and a polished black skin, toilette complete, unless for certain ceremonies a further decoration of flowers or down feathers was required.

The principal article of the man's dress was called waywah. It was a belt, about six inches wide, made of twisted sinews and hair, with four tufts about eighteen inches long hanging back and front and at each side from it, made of narrow strips of kangaroo or paddy melon skins.

For warmth in winter they would wrap themselves in their opossum-skin rugs. Sometimes both sexes adorned themselves with strings of kangaroo teeth fixed into gum, in which a little hole was made, round their heads and necks--yumbean they called them; or forehead bands with hanging kangaroo teeth, which were called gnooloogail.

Pine gum they rolled into small egg-shaped balls, warmed them and stuck them in dozens all over their heads, where they would be left until they wore off, hairdressings being only an occasional duty. The gum they used for sticking the kangaroo's teeth was that of the Mubboo, or beefwood tree.

Sometimes wongins were worn; they consisted of cords round the neck and under the arms, crossing the chest with a shell pendant at the centre of the cross. A shell is still a most prized ornament.

The corroboree dress is one of paint; the feature of it being its design, a man can gain quite a tribal reputation for being an originator of decorative designs.

Their original paint colourings were white, red, and yellow; occasionally they said they got some sort of blue by barter, but very occasionally, as it came from very far. White was from Gidya ash, or gypsum; red and yellow, ochre clay; but they also got both red and yellow from burning at a certain stage certain trees, gooroolay for red; the charcoal, instead of being black, having red and yellow tinges. But since the white people came the blue bag has put yellow out of fashion, and raddle is used for the red.

Their opossum rugs used to have designs scratched on the skin sides and also painted patterns, some say tribal marks, others just to look pretty and distinguish each their own.

Feathers tied into little bunches and fastened on to small wooden skewers were stuck upright in the hair at corroborees, also swansdown fluffed in puff balls over the heads.

The Gooumoorh, or corroboree, is a sort of black fellow's opera; as to the musical part, rather, as some one found an oratorio, a thing of high notes and vain repetition.

The stage effects of corroborees are sometimes huge sheets of bark fastened on to poles; these sheets of bark are painted in different designs and colours, something like Moorish embroideries. Sometimes there is a huge imitation of an alligator made of logs plastered over with earth and painted in stripes of different colours, a piece of wood cut open stuck in at one end as a gaping mouth. This alligator corroboree is generally indicative of a Boorah, or initiation ceremony, being near at hand. Sometimes the stage effects are high painted poles merely.

At the back of the goomboo, or stage, are large fires; in the front, in a semicircle, sit the women as orchestra, and the audience; a fire at each end of the semicircle, as a sort of footlights. The music of the orchestra is made by some beating time on rolled-up opossum rugs, and some clicking two boomerangs together. The time is faultless. The tunes are monotonous, but rhythmical and musical, curiously well suited to the stage and layers. These last have a very weird look as they steal Pout of the thick scrub, out of the darkness, quickly one after another, dancing round the goomboo in time to the music, their grotesquely painted figures and feather-decorated heads lit up by the flickering lights of the fires around.

As the dancing gets faster the singing gets louder, every muscle of the dancers seems strained, and the wonder is the voices do not crack. just as you think they must, the dancing slows again; the voices die away, to swell out once more with renewed vigour when the fires are built up again and again; the same dance is gone through, time after time-one night one dance, or, for that matter, many nights one dance.

The dancers sometimes make dumb-show of hunts, fights, slaughters, the women sometimes translating the actions in the songs; sometimes the words seem to have nothing to do with them, and the dances only a series of steps illustrating nothing.

Corroborees seem to fit in with the indescribable mystery of the bush. That the spirit of the bush is mystery makes it so difficult to describe beyond bald realism, otherwise it seems an effort to seize the intangible. Poor Barcroft Boake got something of the mystery into words.

If an Australian Wagner could be born we might hope for a musical adaptation of corroborees. Wagner was essentially the exponent of folk-lore music, wherein must be expressed the fundamentals of human passion unrefined.

The most celebrated weapon is probably the boomerang the most celebrated kind to whites, though not most useful to blacks, is the Bubberab, or returning boomerang. These are made chiefly of Gidya and Myall. Here these 'Come backs' are never carved, are more curved than the ordinary boomerang, and were greased, rubbed with charred grass, and warmed before being used, so that the slightest warp would be straightened. It is marvellous the accuracy with which an adept can throw one of these weapons, locating it on the exact place to which he wishes it to return.

Gidya is the favourite wood for boomerangs. They are first roughly shaped, then thrown into water and soaked for two or three days; taken out and made into the proper shape, rubbed with charred grass, greased well, and carved in various designs with an opossum's tooth.

Boomerangs have many uses--in peace two clicked together as a musical instrument, as a war weapon, and as a weapon in the chase. Its last and rapidly approaching use will be as a curio for collectors.

Billah, or spears, are made of Belah (swamp oak) or Gidya. These too are cut roughly first and thrown into water, then cut a little more, thrown into water again, and so day after day until finished. Sometimes they are carved with a running feathers-titch-like pattern from end to end, sometimes have bingles, or barbs, cut down one or both sides; some barbarous things with barbs pointing both ways, so that they could be neither pushed out nor drawn through a wound; some are plain, painted at each end or darkened with poison tips.

Billah are war weapons; a larger kind called Moornin are used for spearing emu.

Woggarahs, the hatchet-shaped weapons, were made of Myall, Gidya, and other woods, carved as were boomerangs, each carver usually having a favourite design by which his weapons were recognised.

Booreens, or shields, were of three kinds: a narrow kind made of hardwood, a broad flat kind of Kurrajong, and a medium-sized one of Birah, or whitewood, all painted in coloured designs. It is wonderful the way a man can defend himself single-handed against a number of men, he having only a narrow shield, the only defence he is allowed when he has to stand his trial for a breach of the laws.

Their tomahawks, or Cumbees, were of dark-green stone, of which there is none in this district, so it must have been obtained by barter, as in the first instance were the flat, light Booreens from the Queensland side, and the grass-tree gum from the Narrabri mountains side, for which Gidya boomerangs were given in exchange.

The stone tomahawks have a handle put over one end of the stone, gummed on with beefwood gum, then drawn together under the stone, crossed, and the two ends tied together as a handle, with sinews of emus, opossums, or kangaroos.

Muggils, or stone knives, are just sharpened pieces of stone.

Moorooleh are plain waddies used in war and for killing game; a smaller kind called Boodthul are thrown for amusement.

Boondees are heavy-headed clubs used in war.

The black fellow won't allow his womenkind a heaven of rest, for the spirit women are supposed to make weapons which the wirreenuns journey towards the sunset clouds to get--the women's heaven is in the west--giving in exchange animal food and opossum rugs, no animals being there.

For carrying water they used to make bags of opossum skins. To prepare the skins they would pluck the hair off, and, after cleansing them well, sew up the skins with sinews, leaving only the neck open. They would fill this vessel with air and hang it out to dry.

As, a water vessel, to mix their drinks and medicines in, they used Binguies or Coolamons, a deep, canoe-shaped vessel cut out of solid wood, carved sometimes and painted, a string handle to it. They used little bark vessels to drink out of, like shallow basins, cut from excrescences on eucalyptus trees; these were called wirree. A larger bark vessel they used for holding water, honey, or anything liquid.

While on the subject of personal decoration I forgot the Moobir, or cuts on the bodies, some of which are tribal marks, some marks of mourning, some merely of ornamentation. Both men and women are seen with these marks in the Narran district; some huge wales on the skin from the shoulders half-way down the back, some on the chest and the forepart of the arms. They are cut with a stone knife, licked along by the medicine man, filled in with charcoal, and the skin let grow over.

Various reasons are given for these marks: some say they are to give strength, others as a tribal sign, others just to took pretty. Some give the final reason for everything, 'Because Byamee say so.'

In summer the blacks are great bathers, and play all sorts of games in the water. Their soap is clay; they rub themselves with that, the women plastering it under their arms again and again; the little children rub themselves all over with it, then tumble into the water to wash it off.

In winter they forgo bathing, and rub themselves with liberal applications of grease.

The old blacks used to have very good teeth; they never ate without afterwards rinsing out their mouths, and sometimes munched up charcoal to purify them. But the younger generation have discarded the mouth-rinsing habit, and not yet attained to a tooth-brush: result, gradual deterioration in teeth, a deterioration probably helped by the drinking of hot liquids. Blacks of the old time drank nothing hot. Perhaps, too, their tough meats gave muscular strength to their jaws.

To blacks, kissing is a 'white foolishness,' also handshaking; in olden times even to smell a stranger was considered a risk.

THE AMUSEMENTS OF BLACKS

A VERY favourite game of the old men was skipping--Brambahl, they called it.

They had a long rope, a man at each end to swing it. When it is in full swing in goes the skipper. After skipping in an ordinary way for a few rounds, he begins the variations, which consist, amongst other things, of his taking thorns out of his feet, digging as if for larvæ of ants, digging yams, grinding grass-seed, jumping like a frog, doing a sort of cobbler's dance, striking an attitude as if looking for something in the distance, running out, snatching up a child, and skipping with it in his arms, or lying flat down on the ground, measuring his full length in that position, rising and letting the rope slip under him; the rope going the whole time, of course, never varying in pace nor pausing for any of the variations.

The one who can most successfully vary the performance is victor. Old men of over seventy seemed the best at skipping.

There is great excitement over Bubberah, or come-back boomerang throwing.

Every candidate has a little fire, where, after having rubbed his bubberah with charred grass and fat, he warms it, eyes it up and down to see that it is true, then out he comes, weapon in hand. He looks at the winning spot, and with a scientific flourish of his arm sends his bubberah forth on its circular flight; you would think it was going into the Beyond, when it curves round and comes gyrating back to the given spot. Here again the old ones score.

Wungoolay is another old game.

A number of black fellows arm themselves with a number of spears, or rather pointed sticks, between four and five feet long, called widyu-widyu. Two men take the wungoolays, which are pieces of bark, either squared or roughly rounded, about fifteen inches in diameter. These men go about fifty yards from each other; first one and then another throws the wungoolays, which roll swiftly along the ground past the men with the spears, who are stationed midway between the other two a few yards from the path of the wungoolays, which, as they come rolling rapidly past, the men try to spear with their widyu-widyu; he who hits the most, wins the game. It looks easy enough, but here again the old men scored.

For Gurril Boodthul, if a bush is not at hand, a bushy branch of a tree is stuck up. The men arm themselves with small boodthuls, or miniature waddies, then stand a few feet behind the bush, which varies from five to eight feet or so in height at competitions. They throw their boodthuls in turn; these have to skim through the top of the bush, which seems to give them fresh impetus instead of slackening them. The distance they go beyond is the test of a good thrower; over three hundred yards is not unusual. As practice in this game is kept up, the young men hold their own.

There is another throwing stick somewhat larger than the gurril boodthul, which only weighs about three ounces, and is about a foot in length. The other stick is thrown to touch the ground, then bound on, sometimes making one high long leap, sometimes a series of jumps, as a flat pebble does when thrown along the water in the game children call 'ducks and drakes.'

Yahweerh is a sort of sham trial fight. One man has a bark shield, and he has to defend himself with it from the bark toy boomerangs the others throw. Here again the old men win. Their games, which old and young alike play, are distinctly childish.

Boogalah, or ball, is one. In playing this all of one Dhé, or totem, are partners. The ball, made of sewn-up kangaroo skin, is thrown in the air; whoever catches it goes with his or her division--for women join in this game--into a group in the middle, the other circling round. The ball is thrown in the air, and if one of the circle outside the centre ring catches it, then his side namely, all his totem--go into the middle, the others circling round, and so on. The totem keeping it longest wins.

Goomboobooddoo, or wrestling, is a great Boorah-time entertainment. Family clan against clan. Kubbee against Hippi, and so on. A Hippi, for example, will go into a ring and plant there a mudgee, or painted stick with a bunch of feathers at the top. In will run a Kubbee and try to make off with the stick; Hippi will grapple with him, and a wrestling match comes off. Into the ring will go others of each side wrestling in their turn. The side that finally throws the most men, and gets the mudgee, wins. Before wrestling matches, there is much greasing of bodies to make them slippery.

Wimberoo was a favourite fireside game. A big fire was made of leafy branches. Each player got a dry Coolabah leaf, warmed it until it bent a little, then placed it on two fingers and hit it with one into where the current of air, caused by the flame, caught it and bore it aloft. They all jerked their leaves together, and anxiously watched whose would go the highest. Each watched his leaf descend, caught it, and began again. So on until tired.

Woolbooldarn is an absolutely infantile game. A low, overhanging branch of a tree is chosen, and as many as it will bear, old and young, men and women, straddle it; and, holding on to the higher overhanging branches, they swing up and down with as much spring as they can get out of the branch they are on.

Whagoo is just like our I hide and seek.'

Gooumoorhs, or corroborees, are of course their greatest entertainment, their opera, ballet, and the rest; only they reverse the usual order of things obtaining elsewhere. The women form the orchestra, the men are the dancers, as a rule, though women do on occasions take part too. The dancers rarely sing while performing their evolutions, though they will end up a measure at times with a loud 'Ooh! Ooh!' or 'Wah! Wah!'

There are two dances they think very clever: one a sort of in and out movement with the knees, while keeping the feet close together. Another, which they called I shivering of the chest,' a sort of drawing in and out of their breath, causing a vibratory motion.

Then they give a sort of Sandow performance all in time to the music. They first start the muscles of their legs showing, then the arms, and down the sides of the chest. I am afraid I was not educated up to be appreciative of any of these special wonders, though Matah and others said their muscular training was marvellous.

From a spectacular point of view I thought much more interesting a corroboree illustrating the coming of the first steamer up the Barwon.

The steamer was made--for the corroboree, I mean--of logs with mud layered over them, painted up, a hollow log for a funnel in the middle. There was a little opening in the far side of the steamer in which a fire was made, the smoke issuing through the hollow log in the most realistic fashion. The blacks who first came on the stage were all supposed to represent various birds disturbed by this strange sight--cranes, pelicans, black swans, and ducks. The peculiarities of each bird were well imitated; and as each section in turn was startled, their cries were realistically given. Hearing which, on the scene came some armed black fellows, who, seeing what the birds had seen, started back in astonishment, seemed to have a great dumb-show palaver, then one by one, clutching their weapons, they came forward to more closely examine the new 'debbil debbil.' Here some one would stoke the fire, out would belch through the funnel a big smoke and a lapping flame, away went the blacks into the bush as if too terrified to stay. But you can't describe a corroboree, it wants the scenic effects of the grim bush: tapering, dark Belahs, Coolabahs contorted into quaint shapes and excrescences by extremes of flood and drought, and their grotesqueness lit up by the flickering fires, until the trees themselves look like demons of the night, and the painted black fellows their attendant spirits stealing into the firelight from what seems a vast, dark, unknown Beyond.

The sing-song seems to suit it, and the well-timed clicking of the boomerangs and thudding of the rolled-up rugs. The blacks are great patrons of art, and encourage native talent in the most praiseworthy way; although, judging from one of their legends, you might think they were not.

This legend tells how Goolahwilleel had the soul of an artist, and when his family sent him out to hunt their daily dinner, he forgot his quest and perfected his art, which was the modelling of a kangaroo in gum. When his work was finished, with the pride of a successful artist he returned for applause.

His family demanded of him meat; he showed his kangaroo.

His masterpiece was unappreciated. Even as did Palissy's--of pottery fame--wife, so did Goolahwilleel's family revile him.

His freedom to wander at will, seeking inspiration and giving it form, was taken from him. He was driven out: daily to slay, that his family might feed, and never again was he let go alone--a crowd of relations went with him!

Figure to yourself what a damper to inspiration must have been that crowd of relations; how it must have slain the artist in Goolahwilleel.

How the old legend repeats itself, and now as then, how often the artist is woman--slain that she by the caterer may live. Surely in the interests of intellect was the prayer made: 'Give us our daily bread.'

Perhaps the old legend of Goolahwilleel was originally told with a moral, and that may be: why black artists are so well treated now.

A maker of new songs or corroborees is always kept well supplied with the luxuries of life; it may be that such an one is a little feared as being supposed to have direct communication with the spirits who teach him his art. A fine frenzy is said to seize some of their poets and playwrights, who, for the time being, are quite under the domination of the spirits--possessed of devils, in fact. When the period of mental incubation is over and the song hatched out, the possessed ones return to their normal condition, the devils are cast out, and the songs are all that remain in evidence that the artist was ever possessed.

Some songs do not require this process of fine frenzy they come along in the course of barter, handed from tribe to tribe.

Ghiribul, or riddles, play a great part in their social life, and he who knows many is much sought after.

Most of these ghiribul are not translatable, being little songs describing the things to be guessed, whose peculiarities the singer acts as he sings--a sort of one-man show, pantomime in miniature, with a riddle running through it.

Some which I will give indicate the nature of others.

What is it that says to the flood-water, 'I am too strong for you; you can not push me back'? Ans. Goodoo, the codfish.

What is it that says, 'You cannot help yourself; you will have to go and let me take your place; you cannot stay when I come'? Ans. The grey hairs in a man's beard to the black ones.

'If a man hide himself so that his wife could not see him, and he wanted her to know where he was, yet had promised not to speak, laugh, cry, sneeze, cough, nor move his hands nor feet, how could he do so?' Ans. Whistle.

'The strongest man cannot stand against me. I can knock him down, yet I do not hurt him. He feels better for my having knocked him down. What am I?' Ans. Sleep.

'I am not water, yet all who are thirsty, seeing me, come toward me to drink, though I am no liquid. What am I?' Ans. Mirage.

'What is it that goes along the creek, across the creek, underneath it, and along it again, and yet has left neither side?' Ans. The yellow-flowering creeping water-weed.

'Here I am, just in front of you. I can't move; but if you kick me, I will knock you down, though I will not move to do it. Who says this?' Ans. A stump that any one falls over.

'You cannot walk without me, yet you grease your body and forget me and let me crack, even though but for me you could neither walk nor run. Who says that?' Ans. A black fellow's feet, which he neglects to grease when doing the rest of his body.

With riddles ends, I think, the list of the blacks' amusements, unless you count fights. The blacks are a bit Celtic in that way; some are real fire-eaters, always spoiling for a row. But in most everyday rows the feelings are more damaged than the bodies.

An old gin in a rage will say more in a given time, without taking breath, than any human being I have ever seen; it is simply physiologically marvellous. From the noise you would think murder at least would result. You listen in dread of a tragedy; you hear the totem and multiplex totems of her opponent being scoffed at, strung out one after another, deadly insult after deadly insult. The insulted returns insult for insult; result, a lively cross fire.

It lulls down; the insults are exhausted, quietude reigns. Some one makes a joke, all are laughing together in amity. From impending tragedy to comedy the work of a few minutes. A mercurial race indeed, but not a forgetful one. A black fellow never forgives a broken promise, and he can cherish a grudge from generation to generation as well as remember a kindness.

Though, when high pitched in quarrels, their voices lose their natural tones, as a rule those of the blacks are remarkably sweet and soft, quite musical; their language noticeable for its freedom from harsh sounds.

BUSH BOGIES AND FINIS

WEEWEEMUL is a big spirit that flies in the air; he takes the bodies of dead people away and eats them. That is why the dead are so closely watched before burial.

Gwaibooyanbooyan is the hairless red devil of the scrubs, who kills and eats any one he meets, unless they are quick enough to get away before he sees them, as one woman of this tribe is said to have done on the Eurahbah ridge. It would really seem as if there were a debbil debbil on that ridge; every boundary rider who lives there takes to drink. I think the red spirit must be rum.

Marahgoo are man-shaped devils, to be recognised by the white swansdown cap they wear, and the red rugs they carry. Red is a great devil's colour amongst blacks some will never wear it on that account.

These Marahgoo always have with them a mysterious drink, which they offer to any one they meet. It is like drinking dirt, and makes the drinker dream dreams and see visions, in which he is taken down to the underground spirit-world of the Marahgoo, where anything he wishes for appears at once. The entrance to this world is said to be near a never-drying waterhole, in a huge scrub, near Pilliga. If a man drinks the draught, unless he is made Marahgoo, he dies.

Each totem is warned by its bird sub-totems of the coming of Marahgoo, and after such a warning tribes take care, if wise, to stay in camp; or should a man go out, he will smear his face with black, and put rings of black round his wrists and ankles, and probably have a little charm song sung over him.

Birrahmulgerhyerh are blacks with devils in them, who, armed with bags full of poison-sticks, or bones-called gooweera-are invisible to all but wirreenuns or wizards. Others are warned of their coming by hearing the rattle of the gooweeras knocking together. When the Birrahmulgerhyerh are about, all are warned not to carry firesticks, which at other times after dark they are never without in order to scare off spirits, but now such a light would show the Birrahmulgerhyerh where to point their gooweeras. They are said only to point these poison-sticks at law-breakers, and even then only against persons in a strange country. Their own land is down Brewarrina way, but there they make no punitive expeditions, travelling up the Narran and elsewhere for that purpose.

The Euloowayi, or long-nailed devils, are spirits which live where the sun sets. just as the afterglow dies in the sky, they come out victim-hunting. These Euloowayi demand a tribute of young black men from the camp, to recoup their own ranks.

When this tribute has to be paid, the old men get some ten or so young ones, and march them off to a Minggah at about ten or fifteen miles from the camp. There they make them climb into the Ming-ah, to sit there all day. They must not move, not even so much as wink an eyelid. At night time they are allowed to come down, and are given some meat, which they must eat raw.

The old men from the camp go back leaving their victims with the Euloowayi, who keep the boys up the tree for some days, bringing them raw meat at night. At last they say:

'Come and try if your nails are long and strong enough. See who can best tear this bark off with them.'

They all try, and if all are equally good, the old Euloowayi say:

'You are right. How do you feel?'

'Strong,' they answer.

They are kept on the tree about a month, then taken into the bush to hunt human beings, to deceive whom they take new forms at times. A couple of blacks may be hunting-One will be after honey, another after opossums. The one after opossums will go to a tree, see an opossum, chop into the tree, seize the opossum by the tail as usual. He cannot move him. He'll seize him by the hind legs, still he cannot move him. Then he will hear a voice say, 'Leave him alone, you can't move him.'

The hunter will look down, see nothing but a rainbow at the foot of the tree. Wonderingly he'll come down, and immediately the Euloowayi, who have been in the form of the opossum in the tree and the rainbow on the ground, seize him, tear him open with their long nails, take out all his fat, stuff him up again with grass and leaves, and send him back to the camp. When he reaches there, he starts scolding every one. Probably they guess by his violent words and actions that he is a victim of the Euloowayi. If so, they are careful not to answer him; were they to do so he would drop dead. Any way, he will die that night. When the magpies and butcher-birds sing much it is a sign the Euloowayi are about.

Gineet Gineet, so called from his cry, is the bogy that black children dread. He is a black man who goes about with a goolay or net across his shoulders, into which he pops any children he can steal.

Several waterholes are taboo as bathing-places. They are said to be haunted by Kurreah, which swallow their victims whole, or by Gowargay, the featherless emu, who sucks down in a whirlpool any one who dares to bathe in his holes.

Nahgul is the rejected Gayandil who was found by Byamee too destructive to act as president of the Boorahs.

He principally haunts Boorah grounds. He still has a Boorah gubberrah, a sacred stone, inside him, hence his strength.

He sets string traps for men, touching which they feel ill, and suddenly drop down never to rise again. The wirreenuns know then that Nahgul is about. They find out where he is. Circling, at a good distance, the spot he is on, they corroboree round it. Hearing them, Nahgul comes out. They close in and seize him, kill him, drink his blood, and eat him; by so doing gaining immense additional strength.

Marmbeyah are tree spirits, somewhat akin to the Nats of Burmah. One, a huge, fat spirit--if you can imagine a fat spirit--carried a green boondee, or waddy, with which he tapped people on the backs of their necks: result, heat apoplexy. A few years ago, an old black fellow laid wait for him and 'flattened him out,' since which there has been no heat apoplexy. We think it is because the bad times have made people too poor to overheat themselves with bad spirits of a liquid kind. The blacks differ, and certainly there were some cases of even total abstainers falling victims to the heat wave.

Hatefully frequent devil visitors are those who animate the boolees, or whirlwinds. If these whirl near the house they smother everything with débris and dust.

The Black-but-Comelys say, as they clear the dirt away: 'I wish whoever in this house those boolees are after would go out when they come, not let 'em hunt after 'em here and make this mess.'

The Wurrawilberos chiefly animate these. But sometimes the wirreenuns use whirlwinds as mediums of transit for their Mullee Mullees, or dream spirits, sent in pursuit of some enemy, to capture a woman, or incarnate child spirit; women dread boolees, more even than men, on this account. Great wirreenuns are said to get rid of evil spirits by eating the form in which they appear. I'm sure we all swallowed a good share of the dust devils, but still they came; evidently we were not wizards or witches.

The plain of Weawarra is haunted. Once long ago there was a fight there. Two young warriors but lately married were slain. As their bodies were never recovered, they were supposed to have been stolen and eaten by the enemy. Their young widows spent days searching for them, after the tribe had given up hope of finding them. At last the widows--who had refused to marry again, declaring their husbands yet lived, and that one day they would find them--disappeared.

Time passed; they did not return, so were supposed to be dead too. Then arose the rumour that their ghosts had been seen, and to this day it is said the plain of Weawarra is haunted by them.

Should men camp there at night, these women spirits silently steal into the camp. The men, thinking they are women from some tribe they do not know, speak to them; but silently there they sit, making no answer, and vanish again before the dawn of day, to renew their search night after night.

The high ridges above Warrangilla are haunted by two women, who tradition says were buried alive. Their spirits have never rested, but come out at all times from the huge fissure in the ridges where their bodies were put. Their anguished cries as the stones and earth fell on them are still to be heard echoing through the scrub there; and sometimes it is said one, keener sighted than his fellows, sees their spirit forms flitting through the Budtha bushes, and hears again their tragic cries, as they disappear once more into the fathomless fissure.

There is a tradition--common, I believe, to many black tribes, even outside Australia--that, long before the coming of the white people into this country, two beautiful white girls lived with the blacks. They had long hair to their waists. They were called Bungebah, and were killed as devils by an alien tribe somewhere between Noorahwahgean and Gooroolay. Where their blood was spilled two red-leaved trees have grown, and that place is still haunted by their spirits.

Amid the Cookeran Lake still wanders the woman who arrived late at the big Boorah, having lost her children one by one on the track, arriving at last with only her dead baby in the net at her back. As she died she cursed the tribes who had deserted her, and turned them into trees. Some of the blacks were in groups a little way off; those, too, she cursed, and they were changed into forests of Belah, which look dark and funereal as you drive through them; and the murmuring sound, as the wind wails through their tops, has a very sad sound. She wanders through these forests and round the lake, the dead baby still in the goolay on her back, and sometimes her voice is heard mingling with the voices of the forest; and as the shadows fall, she may be seen flitting past, they say.

Noorahgogo is a very handsome bronze and peacock-blue beetle, said to embody a spirit which always answers the cry of a Noongahburrah in the bush. The bright orange-red fungi on the fallen trees are devils' bread, and should a child touch any he will be spirited away.

Very mournful are the bush nights if you happen to be alone on your verandah. Away on the flat sound the cries of curlews; past flies a night heron; then the discordant voice of a plover is heard. In all these birds are embodied the spirits of men of the past; each has its legend.

Perhaps some passing swans will cry 'Biboh, biboh,' reminding in vain the camp wizards that they too were once men, and long to be again. Poor enchanted swans! to whose enchantment we owe the lovely flannel flowers of New South Wales, and the red epacris bells.

But in spite of their sadness the bush nights are lovely, when the landscapes are glorified by the magic of the moon. Even the gum leaves are transmuted into silver as the moonlight laves them, making the blacks say the leaves laugh, and the shimmer is like a smile.

No wonder trees have such a place in the old religions of the world, and wirreenuns, even as do Buddhists, love to linger beneath their branches--the one holding converse with his spirit friends, the other cultivating the perfect peace.

There would not be much perfect peace about a wirreenun's communing with the spirits if it happened to be in mosquito time. The blacks say a little grey-speckled bird rules the mosquitoes, and calls them from their swamp-homes to attack us. In the mythological days this bird--a woman--was badly treated by a man who translated her sons to the sky; having revenged herself on him, she vowed vengeance on all men, and in the form of the mosquito bird wreaks that vengeance. Her mosquito slaves have just the same spots on their wings as she has.

I dare say little with an air of finality about black people; I have lived too much with them for that. To be positive, you should never spend more than six months in their neighbourhood; in fact, if you want to keep your anthropological ideas quite firm, it is safer to let the blacks remain in inland Australia while you stay a few thousand miles away. Otherwise, your preconceived notions are almost sure to totter to their foundations; and nothing is more annoying than to have elaborately built-up, delightfully logical theories, played ninepins with by an old greybeard of a black, who apparently objects to his beliefs being classified, docketed, and pigeon-holed, until he has had his say.

After all, when we consider their marriage restrictions, their totems, and the rest, what becomes of the freedom of the savage? As with us, as Montague says, 'Our laws of conscience, which we pretend to be derived from Nature, proceed from custom.'

I have often thought the failure of the generality of missionaries lay in the fact that they began at the wrong end. Not recognising the tyranny of custom, though themselves victims to it, they ignore, as a rule, the religion into which the black is born, and by which he lived, in much closer obedience to its laws than we of this latter-day Christendom. It seems to me, if we cannot respect the religion of others we deny our own. If we are powerless to see the theism behind the overlying animism, we argue a strange ignorance of what crept over other faiths, in the way of legends and superstitions quite foreign to the simplicity of the beginnings.

To be a success, a missionary, I think, should--as many do, happily--before he goes out to teach, acquaint himself with the making of the world's religions, and particularly with the one he is going to supplant. He will probably find that elimination of some savageries is all that is required, leaving enough good to form a workable religion understanded of his congregation.

If he ignores their faith, thrusting his own, with its mysteries which puzzle even theologians, upon them, they will be but as whited sepulchres, or, at best, parrots.

GLOSSARY

Bahloo, moon (masculine).

Bibbil, poplar-leaved box-tree. An Eucalyptus.

Byamee, their god; culture hero 'Great One.'

Boorak, initiation ceremony.

Boonal, a sort of flail.

Boobeen, wooden cornet.

Bootha, woman's name; divisional family name.

Boahdee, sister.

Beealahdee, father and mother's sisters' husbands.

Bargie, grandmother on mother's side.

Boothan, last possible child of a woman.

Beewun, motherless girl,

Boomerang, weapon.

Bubberah, a 'come-back' boomerang.

Billah, spear.

Belah, swamp oak.

Booreen, shield.

Birah, whitewood tree.

Boodthul, toy waddy.

Boondee, heavy-headed club.

Binguie, Coolamon; canoe-shaped wooden vessel.

Beewee, brown and yellow iguana.

Bunbul, little boorah ring.

Boormool, shrimps.

Boolooral, a night owl.

Byahmul, a black swan.

Beerwon, bird like a swallow.

Bunnyal, flies.

Binnantayah, big saltbush.

Bohrah, kangaroo.

Boogodoogadah, rainbird.

Buln Buln, green parrot.

Boogahroo, a tree where poison-sticks are kept.

Boondurr, wizard's bag of charms.

Budtha, shrub Eremophila.

Bumble, shrub Capparis mitchelliensis.

Brambahl, skipping.

Boogalah, ball.

Bayarrh, green-head ants.

Bingahwingul, shrub needlebush.

Boondoon, kingfisher.

Bilber, sandhill rat.

Boothagullagulla,, bird like seagull.

Booroorerh, bulrushes.

Burrengeen, peewee; white and black bird.

Bouyoudoorimmillee, grey cranes.

Bouyougah, centipede.

Bubburr, large brown and yellow snake.

Beeargah, crane.

Buggiloo, girl's name; little yam.

Boolee, whirlwind.

Boogurr, things belonging to a dead person.

Bullimah, sky-camp; heaven.

Bulleerul, breath.

Boorboor, come down.

Boyjerh, father, or relation of father.

Brigalow, an acacia.

Birroo Birroo, bird; sand-builders.

Booloon, white crane.

Boonburr, poison tree.

Boorgoolbean, a shrub with creamy flowers.

Birrahlee, baby.

Bahnmul, betrothal of babies.

Boomayahmayahmul, a wood lizard.

Brewarrina, name of place; place of Myall trees.

Boorool, big, great, many.

Birrahgnooloo, woman's name meaning hatchet-faced.

Booloowah two emus.

Bibbilah, belonging to the Bibbil country.

Barahgurree, girl's name; a kind of lizard.

Bogginbinnia, girl's name; a kind of lizard.

Billai, crimson-wing parrot.

Birriebunger, small diver-bird

Burrahwahn, a rat now extinct.

Bralgah, bird; native companion.

Bean, Myall tree; a weeping acacia.

Beebuyer, yellow flowering broom, shrub.

Beeleer, black cockatoo.

Bibbee, woodpecker,

Bullah Bullah, butterfly.

Beeweerh, bony bream.

Buggila, leopard wood.

Bunbundoolooey, a little brown bird.

Brumboorah, boorah song.

Boorahbayyi, boy undergoing initiation.

Boodther, a meeting where presents are exchanged.

Berai Berai, the boys; Orion's sword and belt.

Beereeun, lizard.

Birrahmulgerhyerh, devils with poison-sticks.

Byjerh, expression of surprise.

Buckandee, native cat.

Coolabah, flooded box; Eucalyptus.

Curreequinquin, butcher-bird; piping shrike.

Cumbee, stone tomahawk.

Cocklerina, a rose and yellow crested cockatoo. (Major Mitchell.)

Carbeen, an Eucalyptus.

Collarene, Coolabah blossom.

Cûngil, ugly, nasty, bad.

Cunnumbeillee, woman's name meaning pigweed root.

Dhé, hereditary totem.

Dheal, sacred tree.

Dayoorl, grinding-stone.

Doonburr, grass seed.

Dheelgoolee, a bird-trapping place.

Dardurr, a camp shelter of bark.

Dheala, girl's name.

Dayadee, half-brother.

Dadadee, grandfather on mother's side.

Doore-oothai, a lover.

Dillahga, an elderly man of same totem as person speaking of or to him.

Dooloomai, thunder.

Dillee, treasure bag.

Deenyi, ironbark.

Doowee, any one's dream-spirit.

Dinahgurrerhlowah, death-dealing stone.

Dumerh Dumerh, smallpox.

Dumerh, brown pigeon.

Doolungaiyah, sandhill rat, bilber.

Douyougurrah, earthworms.

Deereeree, willy wagtail.

Durrooee, spirit-bird.

Dinewan, emu.

Dunnia, wattle tree.

Deenbi, diver.

Deegeenboyah, soldier-bird.

Dayahminyah, small carpet snake.

Douyouie, ants.

Dulibah, bald.

Dulleerin, a lizard.

Douran Douran, north wind.

Dunnee Bunbun, a very large green parrot.

Dibbee, sort of sandpiper.

Durrahgeegin, green frog.

Dooroongul, hairy caterpillar.

Durramunga, little boorah.

Doolooboorah, boorah message-stick.

Dulloorah, tree manna-bringing birds.

Eerin, little night owl.

Euloowayi, long-nailed devils.

Euahlayi, name of the Narran tribe.

Euloowirree, rainbow.

Eeramooun, uninitiated boy.

Eleanbah wundah, spirits of the lower world.

Hippi, man's divisional family name

Hippitha, woman's divisional family name.

Inga, crayfish.

Innerah, a woman with a camp of her own.

Illay, hop bush.

Kumbo, man's divisional family name

Kubbee, man's divisional family name

Kubbootha, woman's divisional family name.

Kummean, father's sister.

Kurreah, crocodile.

Kumbuy, sister-in-law.

Kamilaroi, name of a tribe.

Kurrajong, tree; a sterculia.

Moodai, an opossum.

Minggah, spirit tree.

Murrahgul, a bird string trap.

Murree, man's divisional family name.

Matha, woman's divisional family name

Mullayerh, a temporary companion.

Moothie, a friend of childhood in afterlife.

Mirroon, emu net.

Mubboo, beefwood tree.

Myall, a drooping acacia; violet-scented wood.

Moornin, emu spears.

Muggil, stone knife.

Moorooleh, plain waddy.

Moogul, only child.

Mah, hand or totem.

Moograbah, big black and white magpie.

Mirrieh, poligonum.

Mullee Mullee, dream spirit of a wizard.

Mullowil, shadow spirit.

Moolee, death-dealing stone.

Moondoo, wasps.

Murgahmuggui, spider.

Mayamah, stones.

Munggheewurraywurraymul, seagulls.

Matah, corruption of master.

Mooroobeaigunnil, spirits on the sacred mountain.

Midjeer, an acacia.

Mulga, an acacia.

Mooregoo Mooregoo, black ibis.

Moolowerh, a shrub with cream coloured flowers.

Muddurwerderh, west wind.

Mungghee, mussels.

Millanboo, the first again.

Moobil, stomach.

Mouyerh, bone through nose.

Moonaibaraban, spirit sister-in-law.

Mayamerh, Gayandi's camp.

Mullyan, eagle-hawk.

Mirriehburrah, belonging to poligonum country.

Millan, small water yam.

Mooregoo, swamp oak; belah,

Mouyi, white cockatoo.

Maira, a paddy melon.

Mouninguggahgul, mosquito bird.

Maira, wild currant bush.

Mungoongarlee, Largest iguana.

Mooregoo, mopoke.

Mounin, mosquito.

Mungahran, hawk.

Mien, dingo.

Munthdeegun, man in charge of initiate at boorah.

Meamei, the girls; Pleiades.

Mayrah, wind.

Marahgoo, man-shaped devil.

Marmbeyah, tree spirits.

Moorilla, pebbly ridge.

Mahmee, old woman.

Nimmaylee, girl's name; young porcupine.

Nurragah, an exclamation of pity.

Noongah, Kurrajong.

Numbardee, mother and mother's sisters.

Niune, wild melon.

Noongahburrah, belonging to the country of the Noongah.

Noorumbah, hereditary bunting ground.

Noodul Noodul, whistling duck.

Nummaybirrah, wild grape; Namoi.

Narahdarn, bat.

Noorunglely, a setting emu.

Nahgul, a devil haunting boorah grounds.

Oganahbayah, a small eagle-hawk.

Ooboon, blue-tongued lizard.

Oobi Oobi, sacred mountain.

Oonahgnai, give to me.

Oonahgnoo, give to her or him.

Oonahmillangoo, give to one.

Oogowahdee goobelaygoo, flood to swim against.

Oogle oogle, four emus.

Oonaywah, black diver.

Ouyan, curlew.

Piggiebillah, porcupine.

Quarrian, yellow and red breasted grey parrot.

Tuckandee, a young man of the same totem reckoned a kind of brother.

Tekel barain, large white amaryllis.

Tekkul, hair.

Talingerh, native fuchsia.

Tucki, a kind of bream.

Wirreenun, medicine man, wizard.

Wunnarl, food taboo.

Wirreebeeun, young woman.

Wirree, canoe-shaped bark vessel for drinking from, or holding things in.

Wambaneah, full brother.

Wulgundee, uncle's wife.

Woormerh, a boorah boy messenger.

Waywah, man's belt.

Wongin, a string breastplate.

Wogarrah, hatchet-shaped weapon,

Wi, clever.

Weedah, bower-bird.

Wundah, white devil.

Wi-mouyan, magic stick.

Wungoolay, a game with discs and spears.

Widyu Widyu, toy-spear.

Wahl, no.

Wa-ah, shells.

Woggoon, scrub turkey.

Wimberoo, game with leaf and fire.

Woolbooldarn, game; riding on bent branch.

Whagoo, game; hide-and-seek.

Wahn, crow.

Wurrawilberoo, the whirlwind devils.

Waddahgudjaelwon, a birth-presiding spirit.

Wahl nunnoomahdayer, do not steal

Wahl goonundoo, no water.

Weedegah, bachelor's camp.

Wirádjuri, name of a tribe.

Waggestmul, kind of rat.

Wungghee, white night owl.

Willerhderh, north wind.

Wi, small fish.

Wayarah, wild grapes.

Womba, mad, deaf.

Weeweemul, a body-snatching spirit.

Wayambah, turtle.

Yhi, the sun (feminine).

Yarragerh, spring wind, north-east.

Yunbeai, individual totem.

Yarmmara, barley grass.

Yubbil, large bark vessel.

Yungawee, sacred fire.

Yumbean, kangaroo teeth fixed in grim, ornaments.

Yumbui, fatherless boy.

Yaraan, an Eucalyptus.

Yowee, a soul equivalent.

Yahweerh , sham fight.

Youayah, frogs.

Yelgayerdayer deermuldayer, leave all such alone.

Yudthar, feather.

Yubbah, carpet snake.

Yelgidyi, fully initiated young man.

Yowee bulleerul, spirit breath.

ENDNOTES

---

[1] See Mr. Howitt's Native Tribes of South-East Australia, and my Secret Of the Totem, chapter iii.(p. 3)

[2] 'The Beginnings of Religion and Totemism among the Australian Aborigines,' Fortmightly Review, September 1905, p. 452(p. 5).

[3] Ibid. p. 462. (p. 4)

[4] Ibid. p. 454(p. 5)

[5] 'The Beginnings of Religion and Totemism among the Australian Aborigines,' Fortnightly Review, September 1905, p. 452. (p. 6)

[6] Ibid. p. 462. (p. 6)

[7] Ibid. p. 463. (p. 7)

[8] Ibid. p. 465. (p. 8)

[9] 'The Beginnings of Religion and Totemism among the Australian Aborigines,' Fortnightly Review, September 1905, p. 452, Note 1. (p. 10)

[10] For an hypothesis of the origin of the churinga nanja belief, see my Secret of the Totem, chapter iv. (p. 13)

[11] Howitt, Native Tribes of South-East Australia, pp. 57, 467, 694, 769. (p. 21)

[12] R. H. Mathews, J. A. I., vol. xxxiv. p. 284. (p. 24)

[13] Howitt, Native Tribes of South-East Australia, p. 493. (p. 24)

[14] Ecclesiastical Institutions, p. 674. (p. 25)

[15] Waitz, Anthropologie der Natur-Völker, vol. vi. pp. 796-798. Leipzig, 1872. (p. 25)

[16] Journal, Anthropological Institute, vol. xxi. p. 292 et seq. (p. 26)

[17] Magic and Religion, p. 25 sq. Myth, Ritual, and Religion, vol. ii. chap. xii., 1899. (p. 25)

[18] Man, 1905, No. 28. (p. 25)

[19] Observations an the Colonies of New South Wales and Van Dieman's Land, p. 147. (p. 25)

[20] Howitt, Native Tribes of South-East Australia, pp. 488-508. (p. 26)

[21] Howitt, Native Tribes of South-East Australia, p. 500. (p. 30)

[22] Ibid. p. 489(p. 30)

[23] Native Tribes of South-East Australia, pp. 120, 490. (p. 30)

[24] Ibid. p. 494(p. 30)

[25] Journal, Anthropological Institute, XXV., p. 297. (p. 30)

[26] Howitt, Native Tribes of South-East Australia, pp. 121, 125, 453, 455. (p. 49)

[27] Man (1904), No. 53, p. 85. (p. 53)

[28] Here we see the usual antagonism of crow and eagle-hawk.--A. L. (p. 120)